Understanding Living Systems

Life is definitively purposive and creative. Organisms use genes in controlling their destiny. This book presents a paradigm shift in understanding living systems. The genome is not a code, blueprint or set of instructions. It is a tool orchestrated by the system. This book shows that gene-centrism misrepresents what genes are and how they are used by living systems. It demonstrates how organisms make choices, influencing their behaviour, their development and evolution, and act as agents of natural selection. It presents a novel approach to fundamental philosophical and cultural issues, such as free will. Reading this book will make you see life in a new light, as a marvellous phenomenon, and in some sense a triumph of evolution. We are not in our genes, our genes are in us.

Raymond Noble is Honorary Associate Professor at the Institute for Women's Health, University College London. He is a Fellow of the Royal Society of Biology and a chartered biologist, writing extensively on biological theory and philosophy, working on how organisms sense their environment. He held a Rockefeller Senior Research Fellowship with a joint appointment at University College London, where he became Deputy Dean of Life Sciences and Graduate Tutor in Women's Health.

Denis Noble is Emeritus Professor of Cardiovascular Physiology, Department of Physiology, Anatomy and Genetics, University of Oxford. He is a Fellow of the Royal Society and was honoured by Queen Elizabeth II for his work in science. He discovered how protein channels in cell membranes of the heart automatically generate the electrical rhythm. This work led him to challenge the neo-Darwinist gene-centric theory of evolution.

The **Understanding Life** series is for anyone wanting an engaging and concise way into a key biological topic. Offering a multidisciplinary perspective, these accessible guides address common misconceptions and misunderstandings in a thoughtful way to help stimulate debate and encourage a more in-depth understanding. Written by leading thinkers in each field, these books are for anyone wanting an expert overview that will enable clearer thinking on each topic.

Series Editor: Kostas Kampourakis https://kampourakis.com

Published titles:

Understanding Evolution	Kostas Kampourakis	9781108746083
Understanding Coronavirus	Raul Rabadan	9781108826716
Understanding Development	Alessandro Minelli	9781108799232
Understanding Evo-Devo	Wallace Arthur	9781108819466
Understanding Genes	Kostas Kampourakis	9781108812825
Understanding DNA Ancestry	Sheldon Krimsky	9781108816038
Understanding Intelligence	Ken Richardson	9781108940368
Understanding Metaphors in the Life Sciences	Andrew S. Reynolds	9781108940498
Understanding Cancer	Robin Hesketh	9781009005999
Understanding How Science Explains the World	Kevin McCain	9781108995504
Understanding Race	Rob DeSalle and Ian Tattersall	9781009055581
Understanding Human Evolution	Ian Tattersall	9781009101998
Understanding Human Metabolism	Keith N. Frayn	9781009108522
Understanding Fertility	Gab Kovacs	9781009054164
Understanding Forensic DNA	Suzanne Bell and John M. Butler	9781009044011
Understanding Natural Selection	Michael Ruse	9781009088329
Understanding Life in the Universe	Wallace Arthur	9781009207324
Understanding Species	John S. Wilkins	9781108987196

Understanding Living Systems

RAYMOND NOBLE
University College London

DENIS NOBLE
University of Oxford

CAMBRIDGE
UNIVERSITY PRESS

Shaftesbury Road, Cambridge CB2 8EA, United Kingdom

One Liberty Plaza, 20th Floor, New York, NY 10006, USA

477 Williamstown Road, Port Melbourne, VIC 3207, Australia

314–321, 3rd Floor, Plot 3, Splendor Forum, Jasola District Centre,
New Delhi – 110025, India

103 Penang Road, #05–06/07, Visioncrest Commercial, Singapore 238467

Cambridge University Press is part of Cambridge University Press & Assessment,
a department of the University of Cambridge.

We share the University's mission to contribute to society through the pursuit of
education, learning and research at the highest international levels of excellence.

www.cambridge.org
Information on this title: www.cambridge.org/9781009277365

DOI: 10.1017/9781009277396

First published 2023

A catalogue record for this publication is available from the British Library.

*A Cataloging-in-Publication data record for this book is available from the Library of
Congress.*

ISBN 978-1-009-27736-5 Paperback

'This spirited, delightfully readable and accessible refutation of gene-centred biological orthodoxy offers a convincing account of living organisms as active agents and living systems, creatively shaping and responding and adapting to their environments. The authors propose that life resides in the purpose and creativity of the whole organism. Living organisms are not their genes, nor are they determined by – or reducible to – their genes. Instead, genes are tools that the organism actively adapts to further the ends chosen by the organism itself. Written for the non-specialist, while founded on decades of highly respected academic research, the authors' systems approach to our understanding of living organisms heralds a welcome return to common sense and an urgent resetting of our relationship to the natural world in the face of imminent environmental collapse.'

Pauline Phemister, Professor of History of Philosophy, University of Edinburgh, UK

'Ray and Denis Noble have written a wonderful book ... *Understanding Living Systems* is fundamental – for biology and medicine, and trying to overcome our current self-induced environmental crisis. Why? Because their whole argument is based on general systems theory thinking – which they present in an entirely natural, didactic, almost anecdotal style, including a remarkable selection of examples, both real and in the form of thought experiments, to make a series of crucial points. Thereby, the Nobles debunk a series of pernicious myths about how living systems work and evolve, including the self-ish-gene metaphor, gene-centrism, "nature red in tooth and claw". Instead, they celebrate the creativity, synergy, intelligence and agency of living organisms in shaping their own evolution, and that of the endlessly changing, interactive biosphere. This book is a gift to the world – but we can only hope the world will listen.'

Dick Vane-Wright, Natural History Museum, London, UK

'Focusing on the purposive nature of living organisms, Noble and Noble present a powerful and informed view of biology based on current knowledge. They show, with many examples and clear explanations, how a gene-centered view of the world engendered profound misunderstandings about genetics, evolution and ecology, leading to many of the short-sighted and dangerous practices and ideas that underlie current ecological disasters and widespread existential despair. Against the cynical view of narrow self-interest as the engine of life, they describe an integrated and compassionate view of nature based on our best current understanding of biology. Beautifully written, the book can be appreciated and understood by the young generation of scientists, politicians, economists, sociologists and philosophers who are facing the great challenge of acknowledging our misunderstandings, remedying our mistakes and reshaping our world.'

Eva Jablonka, Professor Emeritus, Cohn Institute for the History and Philosophy of Science and Ideas, Tel Aviv University, Israel

'*Understanding Living Systems* is a remarkable achievement. Focusing on the complex *systems* of which DNA is merely one small part, Raymond Noble and Denis Noble convincingly argue that the active *agency* of living organisms plays a central role in both evolution and development. In this wonderful book, the authors meticulously present a perspective that offers an understanding of life that touches on events at molecular levels, cultural levels, and all of the analytical levels in between. From their deep understanding of what actually happens inside the living cells that constitute our bodies, Raymond Noble and Denis Noble ascend to a great height, offering a breathtaking view of what it means to be an intelligent animal embedded in sociocultural contexts, embodied within complex ecosystems, and animated by purpose.'

David S. Moore, Professor of Psychology, Pitzer College and Claremont Graduate University, California, USA

'Takes the story of evolution from where Darwin left it, including his ideas on creative purpose and acquired characteristics. By adding the control of chance by organisms, it makes Darwin's theories compatible with the freedom to choose. Elegantly and clearly written, at the hopeful core truth of our lives.'

Samuel Shem, Professor of Medical Humanities at NYU School of Medicine, USA; author of *The House of God*, and *The Spirit of the Place*

'The Noble brothers have done an enormous service to biological understanding of evolution in this short book.'

Anthony Trewavas, Professor Emeritus, University of Edinburgh, UK

Contents

Foreword

Are we slaves to our genes? Are genes selfish replicators that use us as vehicles? Do organisms have agency and purposes? These are just a few of the many questions addressed in this fascinating book by Ray Noble and Denis Noble. In a unique, literary prose the authors begin their foray into living systems from what is often considered as the ultimate organiser of life: the genome. In contrast to this reductionist view, the authors show that genomes are important only in their own context and only when we consider their interactions with that context. Most importantly, as we move from the genome to 'higher' and more complex levels of organisation – to cells, tissues, organisms, ecosystems – what we see is the purposeful action of organisms and multiple, constant interactions across all levels. If you believe that our genes and genomes are our inner essences that somehow determine who we are and what we do, then the present book is likely to change your mind forever. Ray Noble and Denis Noble have produced the most concise and readable account of how living systems operate, interact and produce the unique phenomenon we call life. Life is more than a code written in DNA; it is more than chemical interactions; life is the continuous operation of complex systems, with interactions at all, and across all, levels. Reading this book will make you see life in a new light, as a marvellous phenomenon, and in some sense a triumph of evolution.

Kostas Kampourakis, Series Editor

Preface

This book addresses four fundamental misunderstandings about living organisms. In the first half of the twentieth century, and particularly in the late 1960s and early 1970s, ideas on living organisms and their evolution were formulated in what became known as the Modern Synthesis. For this there were four main pillars: (1) that changes in the structure and function of organisms in one generation could not be passed on through the germ line – this dogma was formulated by August Weissman in 1883 and, in the mid-twentieth century, became a fundamental part of a gene-centric dogma; (2) that organisms could not alter their genes, so causation was held to be a one-way process, from gene to organism functionality; (3) that the organism was best viewed as a passive vehicle for retaining genes in a 'gene pool' and, most significantly, that the behaviour and function of organisms was controlled to this end (this gave birth to the selfish-gene concept, popularised by Richard Dawkins in his bestselling book, *The Selfish Gene*); (4) that evolution occurs through small random changes in genes (gene mutations) that are passively selected in the process of natural selection. What we show in this book is that none of these pillars is correct, or stands as originally formulated. It is clear now that selective changes in the organism can be passed on through the germ line; that there is no hard barrier to the germ line, and that some key factors, such as epigenetic changes, do pass through that; and that there can be no 'gene pool' that is separate from the organisms. So the vehicle–gene separation is simply mistaken, both philosophically and scientifically.

The Modern Synthesis led to the view that organisms are passive living systems, which experience environmental changes but play no active role in the process of evolution. On the contrary, we show that organisms are active agents in evolution. This book also addresses another fundamental misconception: that DNA is a 'secret code' or set of instructions to the organisms. Furthermore, the evolution of species has involved major rearrangements of the genome, not just small random mutations. Natural selection is also an active process performed by organisms, not a passive one. Most of what we do does not involve genes directly, and that is particularly true of our behaviour. We do not have a gene for selfishness, and if we did, we would also have a gene for altruism; but these behaviours are more complex than can be explained by a gene hypothesis. Genes are not directly engaged in our behaviours. The only way we have been able to understand genes in relation to function, health and wellbeing is through large-population association studies; that is, the relationship is probabilistic rather than determinate. We discuss why and how this restores agency to organisms.

Understanding living systems involves understanding their agency.

1 The Gene Delusion

The view of living systems as machines is based on the idea of a fixed sequence of cause and effect: from genotype to phenotype, from genes to proteins and to life functions. This idea became the Central Dogma: the genotype maps to the phenotype in a one-way causative fashion, making us prisoners of our genes.

How did this dogma become entrenched? One of the great achievements of twentieth-century science was the discovery of the structure of genetic material, DNA (deoxyribonucleic acid). It heralded a new era of biological understanding, but it also created a gene-centric view, the gene as a 'code' or 'blueprint' for life. By unravelling the 'code' we could find the ultimate or 'primary' cause of life and its functions. It led to the new field of genomics, which would seek to associate particular genes, or groups of genes, with particular functions. But if all functions could be reduced to genes, then this produces a problem: where lies the agency of organisms? Could this also be reduced to genes? Are we driven by genes? The answer presented by those central to what came to be called the Modern Synthesis was that it could – even to the point of arguing that our freedom to act (free will) and our agency is an illusion.

This dogma is a distorted view because it separates genes, as replicators, from the organism, as a vehicle. Genes are seen as both the driving force and the goal of organisms. Yet, while genes are essential in making proteins, an ability that must be passed on through the generations, they are part of a regulatory system and not its directors. The director is the self-regulating organism. The organism does not wait for commands given by genes. Just as musical notes can be arranged in many ways to create compositions, organisms use their genetic heritage, implementing a diverse range of possible outcomes. Furthermore, when their heritage is inadequate to cope with environmental stress, organisms can alter their genes. Thus, organisms can change their genetic heritage.

What Is a Gene?

There are two key misconceptions about genes, concerning (1) what they are and (2) what they do. In common language when we talk today about genes we tend to think of DNA, codes, a book of life or a blueprint. But do we really understand what we mean by genes? We talk about genes as though we know what they are. Genetics is a major subject, along with genomics. Students of biology learn about Mendelian inheritance. Mendel discovered that inheritance could be viewed as discrete, and that there exist dominant and recessive forms of each discrete inherited characteristic. He showed this by demonstrating the probabilistic nature of inheritance. Different characters can be determined by whether a gene is said to be dominant or recessive. But the gene of Mendelian genetics is very different from the concept used by those who study genes today. Significantly, it is different from that used by Richard Dawkins in his book *The Selfish Gene*, which changes from context to context. In fact, the gene as a concept has become slippery, like a conjurer's sleight of hand. One moment it is an independent inherited characteristic; the next, it is a DNA sequence. But these are not identical, and confounding the two creates a conceptual muddle.

An example of gene confusion was shown in a debate in 2009 between Richard Dawkins and Lynn Margulis. Margulis, who demonstrated symbiogenesis (the coming together of two species to form a new one) as a significant step in evolution, challenged Dawkins by saying that there is more involved in the inheritance of a character than DNA.

'I would embrace that gladly as a new "honorary" gene. That's fine,' Dawkins quickly replied as the audience groaned.

'Why not, why not?' Dawkins insisted.

So, in that sense, Dawkins' trick is that a gene is anything and everything that is inherited. Yet the thesis of *The Selfish Gene* requires genes to be discrete entities whose frequencies can be measured.

Can a Gene Be Selfish?

You might think this is simply a problem of semantics, but it is important in understanding what genes do. A gene cannot be selfish if it is simply part of something else that is the purposive entity – the organism. Only the purposive entity could be considered selfish. Genes do not and cannot make choices; organisms can and do. You might also think that 'selfish gene' is simply a colourful metaphor. If that is so, then it is a powerful misrepresentation of the concept of selfishness, which can only be attributed to wilful beings performing deliberate acts. Yet it is a seductive argument: we are selfish because of our genes, since we exist to pass genes from one generation to another. As Dawkins writes, the genes are 'manipulating it [the organism] by remote control'.

Slippery or not, genes are insufficient to account for development and function. Genes alone are not responsive to changing circumstances. Their expression is a function of the system as a whole, not of the gene itself. Genes are not 'swarming within us', controlling our function, and certainly not on a moment-to-moment basis. On the contrary, they are highly regulated; else, we as organisms would have no coherence. So, just as the paint on an artist's pallet cannot alone form a picture, so genes cannot create an organism.

Genes Do Not Make Choices

In this sense of being used by the system, genes are not agents in causality; they are templates, tools enabling the organism to develop and function. Indeed, much of what is necessary for organisms to function is not inherited at all. It develops with the organism and might require learning. The functional development of our visual system, for example, requires experience of our visible world. The number of synapses (the tiny functional connections between nerve cells) in the visual cortex per hemisphere is around 32 billion in rats and a staggeringly large number in humans. These are honed after birth and during early life. The complex connections of our brains continue developing into our late twenties. So interactions with the environment are involved in guiding and organising these complex functional anatomical arrangements. Genes alone are not sufficient.

Similarly, our social functioning requires interactions as social beings. The amygdala, for example, is an area of the brain that is thought to attach emotional value to faces, enabling us to recognise expressions such as fear and trustworthiness, while the posterior superior temporal sulcus predicts the end point of the complex actions created when we as agents act upon the world. That is, we anticipate the world, and in large part this requires learning. These complex interactions are not governed by genes, which is why early-life experiences may have such critical effects upon us – on our emotions and character traits.

A False Dichotomy

Even if we could clearly define genes as discrete entities, their contribution to our functionality cannot be understood apart from organisms or their habitats and life experiences. So, for a large part of our lives, there is no need for genes other than in some of our capacities to act (see Chapter 4). Most of our decisions as humans do not engage with genes directly at all. If we decide to be kind to someone, we do not engage a kindness gene; if we are cruel, we do not engage a cruelty gene. Even entirely unconscious functions such as our heartbeats, so essential to life, do not directly involve genes.

Early Transitions in Evolution

One of the major transitions in the evolution of life was the cooperation of microscopic cellular organisms to form multicellular organisms. They can range from simple aggregations of many similar unicellular organisms into colonies, as for example in *Volvox*, the globe algae that can form spherical colonies of as many as 50,000 cells. At the other extreme are organisms like us with many colonies of separately specialised cells, arranged as individual organs and tissues.

That change to multicellularity enabled living organisms to become much larger. But it also created a problem. Cells need to access food and respiratory gases. Single-cell organisms can do that by exchanging material directly with the watery environment in which they grow and divide. The molecules simply diffuse within the water. However, there is a limit to the distance over which this can occur sufficiently quickly, and that limit is microscopic. It is around 50 micrometres (μm), which is only about one-twentieth of a millimetre, about the thickness of a hair. Incredibly, all the cells in our bodies are bathed in moving fluids that get this close to the cells. This is the circulatory system, and at its centre lies a pump, the heart, while out in the rest of the body it branches into a fine meshwork of capillaries. They bathe all the cells of our bodies sufficiently rapidly that they can all take up nutrients and oxygen and pass back carbon dioxide and other molecules they need to get rid of.

The First Organ

Something needs to work to push the fluids around the body, and it needs to develop as the first functioning organ of the body. The embryo cannot grow beyond a very tiny mass of cells unless this happens. Long before any other organs and body shapes such as limbs and digits form, a tiny tube of cells starts beating. In humans, this may be as early as 3–4 weeks after conception. But you would not be able to hear this heartbeat until a little later (it can be picked up on hand-held ultrasound or stethoscope at around 8 weeks).

How does it do that? The heartbeat is extremely robust. In a typical human lifetime, it will continue for 3 billion times over a period of 70 or more years. Like other critical life processes, it has therefore to be very robust.

Genes and the Heartbeat

What role do genes, as DNA sequences, play in this critically important process? The answer is that they enable the essential proteins for heart rhythm to be made. They provide the templates from which all the types of proteins are made, which then sit in the cell membrane and enable ions (electrically charged atoms) to cross the heart cell membrane. It is through these channels that tiny electrical currents flow to initiate the heartbeat; and it is this electrical activity that is picked up in an electrocardiogram (ECG). But what controls this process? The answer is, the heart cell itself. The cells literally tell the genome how much of each type to make. The genes do not themselves generate heart rhythm.

The reasons for that are interesting. There are ways in which a DNA sequence could be involved in a cyclical process that generates rhythm. The daily rhythm we call circadian sometimes does involve a causal sequence that involves a DNA sequence.

But heart rhythm could not possibly do that. It is far too fast. It takes tens of minutes or even hours for the production of protein to occur after the activation of a gene. The heartbeat is usually faster than 1 per second. There is no time for any change in the production of particular proteins.

What happens therefore is that the proteins and their interactions with each other, membranes, and many other chemicals somehow generate rhythm. How they do this is fascinating.

The story has so far developed over around 60 years of research. In 1960 one of the authors (DN) was working with his thesis supervisor on a very few of the proteins that form ion channels in the heart. At that time only four channels were known: a rapidly activated sodium channel, two potassium channels, one of them very slowly activated, and an anion (a negative ion) channel. But this was enough to show a very important property of cells. The interactions between the proteins, the membranes and the networks within which they find themselves can automatically produce a rhythm as important as the heartbeat. They do so by using a universal electrical property of cells. Their membranes and proteins can generate and maintain large electrical differences between their interiors and their exteriors. Expressed as a voltage, it is small, typically lying between minus 100 mV (only 0.1 volts) and about plus 50 mV (0.05 volts). But the membrane is extremely thin, so these voltage fields can be as large as 30 million volts per metre. That is a huge field strength, equivalent to a bolt of lightning.

Now we come to an important consequence of the strength of that field. It can cause proteins to change shape. In the case of ion channel proteins, the shape change can be the difference between the channel being open and being closed.

When they are open, the channels themselves cause changes in the voltage, since they carry charged particles into or out of the cell. So the proteins and the cell membrane form an automatic feedback loop (Figure 1.1). A change in protein shape can cause a change in field, which in turn causes the channel to open or close. Some channels open when the cell electrical potential becomes positive, others close.

What was shown 60 years ago is that this feedback loop is sufficient to produce electrical rhythm. No single protein, nor the gene forming its template, can do that alone. The interaction is the key. It can and does generate rhythm, and at about the same frequency as natural heart rhythm.

So was the problem of heart rhythm solved 60 years ago? Yes and no. Yes, because it was possible to show mathematically that rhythm of the right frequency could be generated by this mechanism (Figure 1.2). No, because

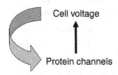

Cell voltage

Protein channels

Figure 1.1 Feedback loop between cell voltage and protein channels. Changes in cell voltage open and close protein channels, which then change cell voltage.

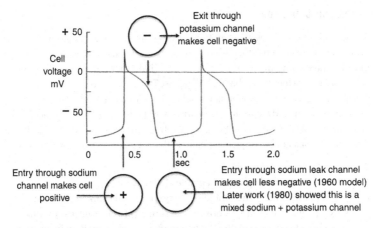

Figure 1.2 The first mathematical model of electrical heart rhythm, based on early experiments on sodium and potassium channels in heart cell membranes. The calculated voltage changes show voltage changes during heart rhythm very similar to that in a real heart. The rhythm is generated by the opening and closing of ion protein channels and results from the feedback interaction (Figure 1.1) between the channels and the cell voltage.

the rhythmic mechanism is very fragile. Knock out any one of those four proteins, or their DNA templates, and the rhythm would suddenly stop.

What subsequent research has shown is that many more than four proteins are involved. So many that, even when one of the key proteins is knocked out or blocked, the rhythm continues.

This is a fundamental property of the networks of interactions in living organisms. They are fail-safe. That robustness depends on the involvement of many genes and their proteins. Even when one is absent, the networks still function well. This is the fundamental reason why association between any one gene and function in the organism has been found to be exceptionally low. It is also a reason why getting your DNA sequenced will usually only tell you a limited amount of information on the diseases you are likely to suffer from. Only if you suffer from a serious but rare genetic disease will the information be useful. But you and your doctor will usually know that anyway from your symptoms and family history.

Genes and Causation

So the problem of genes arises not only from how they are defined, but also from how we view what they do or why they are needed. Most of the time, from moment to moment, life carries on its vital functions without directly using the store of templates that we call genes. Those are used only when more proteins need to be made. Giving causal primacy to what is inherited has led to a profound misunderstanding of organisms and life. Yet modern science has given these poorly defined entities a role both as primary causal agents and as the ultimate measurable objective of life. In that view, we exist to preserve our genes.

The difficulty of providing an acceptable definition of a gene is the key to a surprising proposition. The problem can be put in one simple question: do genes exist? This may seem an absurd question. Yet it is a necessary one – because genes do not exist in the way that is often assumed or thought. In large part the gene is illusory, at least in the sense that there is not a gene for this and a gene for that. Nor are genes directly causal of function. Let us examine this further.

More Is Inherited Than the Genome

If the word gene is taken to mean anything inherited, then we are left with a tautology. It is definitively true that an extended version of 'gene' *must* be anything inherited. This gets us nowhere. The circularity of this argument is apparent. If we end up saying a trait is a gene, then the concept of a gene as a causal factor in a trait is meaningless.

Consider the complexity of arrangements in an economy. The goods you get from a store are not themselves accountable for you purchasing them. Inheritance is a process, not a discrete, measurable entity. What we inherit is a propensity to do things. So we inherit not in two discrete parts, cause (gene) and effect (trait). We inherit a capacity of becoming or being, where cause and effect are one, as in generating heart rhythm. There isn't a gene that determines the price of goods; there is a complex interaction between producers and consumers. Nor is there a gene that determines that you will always buy the cheapest. If you think this is insignificant, consider that your life and how you live are primarily influenced by this interaction. Nor is there a gene or genes for the most significant aspect of our lives, our habitats. It might be said that markets operate on the assumption that we are inherently selfish; but it might equally be the case that we are selfish because of the way markets operate. Behaviourally, functionality is not directly influenced by genes.

Yet we inherit our created environments, for the most part, modifying them and passing them on to generations to come. This includes our homes and the things we use in our lives. The same is true of our culture. We also inherit our ideas and develop them, changing our view of the world in which we live. This is why we write books, plan and work together in solving problems. Where something does not work, it can be discarded or changed. This is not in our genes. This is the nature of the environment within which our behaviour evolves and changes. It is too simple to see us as hunter-gatherers living in a concrete jungle, for our biology changes in relation to our environment. We continually create and are part of our environment. This active creation influences us, body and mind. This is a continuous process of change and adaptation. Understanding how our built environment influences our well-being is important. Just as we see in our hearts, genes are not directing our behaviour.

Genes Are Not Agents

So inheritance and genes are two separate concepts. Adopting too readily the idea of genes as the sole agents of inheritance is a mistake. Genes can have no such agency and furthermore cannot contain all the information that life needs. We should move from 'genes for everything' to a more refreshing view that life requires no such discrete agency. If genes are not agents, perhaps we don't need them, or the gene of the textbook does not exist. So, then, what is a gene? The common view of the gene is that it is a package of information that is used to create something or do something. However, as we have seen, the meaning of the word 'gene' depends on its usage. When challenged with evidence for inheritance independent of DNA, Richard Dawkins replied that the word 'gene' includes anything that is inherited. As we have seen, this is meaningless. Genes in that sense are vague and malleable rather than the discrete functional entities required in the context of the 'selfish gene'. We certainly need tools that enable organisms to build proteins and maintain the fabric of cells, tissues and organs. But this is undoubtedly not Dawkins' gene. So where did the contemporary concept of the gene come from?

The Modern Synthesis

The concept of a gene as an independent hereditary unit was formulated as the Modern Synthesis of evolutionary biology in the 1930s and 1940s. This was not a Darwinian synthesis, since it specifically rejected some of Darwin's key ideas, including the inheritance of acquired characteristics, sexual selection and much else. The objective was to reduce Darwin's ideas to a narrow version of what he actually said. The only synthetic aspect was to incorporate Mendel's work. The assumption is that each gene is responsible for specific characters. It focuses on genes by assuming that characters are determined by genes. It formed the basis of what is now called population genetics, based on calculating the frequencies of ideally independent characters as genes.

With the discovery of the structure and role of DNA in the 1950s, the concept of a gene changed again, producing a specific molecular basis for the gene: DNA sequences, forming templates for proteins. Seeing genes as discrete functional entities of inheritance led to another sleight of hand: conceptually separating the genome from the organism and then regarding organisms as vehicles for passing genes from one generation to the next. Thus, maintaining genes in the population became both a measure of life's success and its ultimate purpose.

A Conceptual Error

Conceptually, genes were given a life of their own in life's narrative, as if they were manipulating organisms to their own ends. Like drivers of cars, they were seen to be controlling the destination. In this view, genes are primary causal agents, and because they are discretely passed from generation to generation, it is the genes themselves that are the subjects of natural selection in evolution. Organisms were reduced to mere vehicles, carrying genes to the next generation, with the survival of genes the primary objective of life. It is a curious fairy tale in which genes are selected, phenotypes are not. The story came with a fantasy land called the 'gene pool'. Yet it is the phenotype that acts in nature.

The 'Gene Pool' Mythology

This gene-centric view is now culturally embedded; so much so that it is difficult to challenge. Thus the myth of genes as causative agents was born, a belief system that traits are determined by genes. Yet counting genes in a 'gene pool' as the objective of life is a meaningless concept. It carries little weight in terms of functionality, just as counting shoelaces tells us little about the function of boots and shoes, other than that they might require different lengths of lace. It is certainly not the purpose of boots to carry shoelaces.

The Dualist Problem

The mechanistic view of life has prevailed and has led to a mistaken view of causality in living systems. René Descartes thought that living organisms were like machines. All animals are automata. This created a fundamental problem that persists to this day, which is that, if this is so, where does our freedom of will to act come from? Descartes' solution to this problem was to suggest that only humans have a soul and could feel pain. This separation of mind and body is now called Cartesian dualism. Humans had the (mechanical) body of an animal, but the freely acting separate soul of a human.

Descartes was both a scientist and a philosopher. In the seventeenth century it was not unusual for scholars to cover both disciplines. In many ways, it is a pity that people now think science and philosophy are separate disciplines. We are taught that scientists study facts while philosophers merely speculate. But the way in which we interpret facts depends on how we see them, in what context, and on what they might mean. We make assumptions about the world, not all of which are directly testable. But it is better that we know this, for it can prevent simple errors of interpretation. Everything we see, hear or feel depends on how we interpret our senses.

This separation of mind and body and the gene-centric view have something in common, which is that they both look for an organiser of something that is in fact self-organising: the organism.

In his treatise on man (*De l'homme*), published posthumously in 1662, Descartes explained his ideas on the embryo:

> If one had a proper knowledge of all the parts of the semen of some species of animal in particular, for example of man, one might be able to deduce the whole form and configuration of each of its members from this alone, by means of entirely mathematical and certain arguments, the complete figure and the conformation of its members. (*On the formation of the fetus,* part of the *Treatise on Man*)

The Gene-Centred Duality

It is a beguiling idea. When we look down a microscope at a cell, we see a distinct nucleus containing a large part of the cell's DNA in the chromosomes. It is easy enough to perceive the nucleus as a central governing structure, sending out its instructions to the cell, a bit like central government in London or Washington DC. But it isn't, and it doesn't. On the contrary, it is controlled by the cell and by the organism.

This idea of a central governing structure is at the origin of the development of a great twentieth-century illusion: the Central Dogma of biology. Descartes expresses the idea very clearly. It is that by knowing all the component parts of the semen one could mathematically predict the development of the embryo and the future adult. The modern version is of course the idea that the development of the embryo is specified by the genome, sometimes itself called 'The Book of Life'.

The Central Dogma and the Human Genome Project make a fundamental error in attributing purpose to genes. However, genes are not alive and can do nothing without the organism. This conceptual separation of the genome from the organism has produced a new kind of dualism where the genome is considered to be the main driving force of all our behaviour, and makes the fundamental error of assuming that organisms only exist to preserve the genome.

2 Replication, Reproduction and Variation

If the dichotomy between replicator and vehicle is wrong, then what is it that replicates? The purpose of reproduction is not replication, at least not exactly. Reproduction brings about change. It shakes up the templates and provides new avenues to explore in adapting to a changing environment. It creates and propagates variation. But it also provides a way for the lessons learned in one generation to be passed on to the next. Reproduction is sensitive to the environment of the parent generation and enables change through the germ line.

Erwin Schrödinger's Mistake

Erwin Schrödinger's seminal book, *What is Life?*, published in 1944, contains two fundamental errors that continue to influence thinking about genetics. The first is that what is passed on to subsequent generations is so vast that it must all be stored at the molecular level. It is indeed vast, which is one of the reasons it is not all stored at the molecular level. The second is that the genetic material faithfully self-replicates like a crystal.

Physicists already knew about the molecular structure of crystals and how such arrangements could replicate automatically by incorporating further similar molecules to form an ever-larger lattice. There are two problems with Schrödinger's idea. First, if the information is stored in a molecular crystal, it wouldn't contain molecular sequences that could be interpreted by the organism unless it was an unusual crystal. Second, the form would be only just enough to enable that crystal structure to grow rather than replicate.

That is how crystals develop from a small beginning. Many a schoolchild has watched this process with amazement, using a chemistry set.

The Replicator Problem

Schrödinger's explanation of replication, however, is where the real problem begins. A standard crystal replicates simply by forming a template for further molecules of the same type to add themselves endlessly to the growing crystal. The genetic material does not do this. Schrödinger would not have known this, but he sowed a seed that influenced the development of the Modern Synthesis in three critical ways:

- The genetic material self-replicates. It does not. It depends on the cell.
- The genetic material replicates faithfully. It does not, and the extent to which it does depends on the cell.
- Faithful replication is the purpose of life. It is not. Responding to change is.

No geneticist today would imagine that DNA replicates like a crystal, but the idea of faithful self-replication persists.

When cells divide, DNA is replicated with amazing accuracy. But this does not occur by accident. It occurs through an equally amazing cellular process. DNA does not achieve this by itself. On its own, there would be hundreds of thousands of mistakes every time the genome is copied. Accurate replication can only be performed in a cell with all the cellular error-correcting machinery to reduce an error rate of 1 in 10,000 to just 1 in 10 billion. The correcting process involves an army of unique proteins (mismatch repair enzymes) organised by the cell. Furthermore, this provides a way for the cell to alter the DNA, because it is a targeted process.

Significantly, Schrödinger realised that there was a problem with his ideas. In his 1944 book, he wrote:

> We seem to have arrived at the ridiculous conclusion that the clue to the understanding of life is that it is based on a pure mechanism, a 'clock-work' ... The conclusion is not ridiculous and is, in my opinion, not entirely wrong, but it has to be taken 'with a very big grain of salt.' (*What is Life?*, pp. 101–102)

He never explained what was in that 'very big grain of salt'. We now know. It concerns the subsequent development of his idea into the Central Dogma of molecular biology.

The Double Helix

Schrödinger's idea significantly influenced Watson and Crick in their approach to the structure of DNA, even though DNA is not a crystal inside a cell. Nevertheless, in *The Selfish Gene*, Dawkins writes that DNA 'replicates like a crystal'. Once seeded, crystals grow automatically. DNA inside a cell does not. Nor can it do so outside a cell. This is fundamentally different from crystallisation, since DNA replication is under the active control of the living cell. Unlike a crystal, the process is not inherent within the DNA itself. The cell controls DNA. The duplication of the strands of DNA is achieved through an orchestrated cellular process, during which many errors will occur. These must be corrected. A section of DNA is unwound and each base bonds to the partner for which it has an affinity: thymine (T) naturally bonds to adenine (A), guanine (G) bonds to cytosine (C). Errors occur because the pairing affinity is not exact. While the strands are still exposed an army of mismatch repair enzymes (DNA polymerases) produced by the cell follows on behind, detecting erroneous pairings and correcting them.

DNA Correction

DNA sequences are thus corrected one by one by cellular mechanisms that would outperform even the best human copy-editors. Imagine a copy-editor working almost word-perfect through the texts of hundreds of books. Nature took a couple of billion years to evolve such a magnificent trick. It is a trick that turns on its head the idea that genes control cells. The organism itself controls this remarkable spectacle. The genes dance to the tune of the cell. But why is this dance necessary?

With a natural error rate in DNA copying, complex organisms could not function. It would produce hundreds of thousands of errors that would destroy the organism. Furthermore, the error-correcting process needs to be able to identify and target the errors. The error-correcting process cannot function without this ability. The double helix structure enables the fitness of two nucleotides to be detected. Without this, an error could not be detected. But the double helix, in itself, does not explain how the corrections happen.

Dancing to the Tune of Life

In organisms, from amoeba to humans, there are always two threads of DNA. The two threads are wound around each other, forming a double helix (Figure 2.1). Each is a mirror image of the other. The base in one line naturally

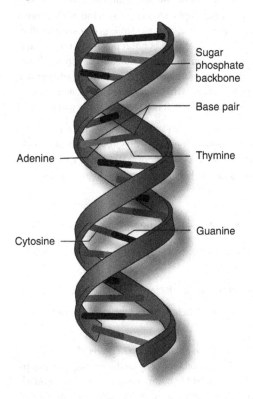

Figure 2.1   Diagram of the DNA double helix.

forms a bond with its partner base in the other. Thus, C and G connect as a couple, as do A and T. A mismatch occurs when this rule is broken. The error-correcting process detects these mismatches and restores the correct base pairing through those specialised proteins (the mismatch repair enzymes) that can cut and paste DNA sequences. These proteins perform this function following the replication process along the thread, picking the errors up as they go along. Many of the errors are flagged, so informing the cell that there is a problem which may lead to the replication process being halted.

So replication is clearly a cellular process. This is not surprising in the growth of organ tissues, because that requires replication of cells. In the production of gametes, however, each gamete in sexual reproduction receives only half of the chromosomes. Reproduction in that case is not replication; on the contrary, it is a mechanism of change involving the shaking up of genes. All this serves to demonstrate that the conceptual separation of replicator and vehicle leads to erroneous assumptions about what natural selection acts upon – because, where the replication takes place, it is a cellular and organ process.

Sexual reproduction is not the only way in which this mixing of DNA can occur. Bacteria, for example, obtain the same benefits by simply freely exchanging segments of DNA. And this process of DNA and RNA exchange still exists in multicellular organisms, because cells exchange vesicles containing nucleotides.

DNA Sequence Triplets – The 'Secret of Life'?

On 28 February 1953, Francis Crick excitedly entered the Eagle pub in Cambridge, where he often had lunch.

'We have discovered the secret of life!' he declared triumphantly.

Watson and Crick had just developed their model of the structure of DNA. It worked.

The discovery of the double helix is the great triumph of molecular biology in the twentieth century, matched in importance only by the finding that groups of three nucleotides (triplets) correspond to particular amino acids, the building blocks of proteins. Both of these discoveries were fantastic achievements.

So what, then, is the problem? The secret, if there is one, lies in how cells use and orchestrate the DNA.

The Genome Is Not the 'Book of Life'

As wonderful as it is, DNA is not a mystical factor controlling the cell. DNA is a tool of the cell. A significant part of the DNA is stored in the nucleus, giving an erroneous impression that this is the centre of the organisation, a bit like municipal offices in a local authority. But cells are not controlled from the centre. So there are two erroneous assumptions about DNA: one, that it is some sort of 'code' or 'blueprint' for life and, two, that it is a central organiser. Both are false. 'Code' is a metaphor, 'organiser' is simply wrong. It is the cell that organises itself. It is the organism that parcels out its DNA in the form of its chromosomes to pass these tools on to the next generation. Furthermore, organisms may share the tools. Perhaps the real secret was that we failed to see this. The focus on DNA as a code may be why this was so.

There is a beauty in the DNA. With just four nucleotides, the cell can manufacture almost all the amino acids it needs. Nearly all our proteins are made with just 20 types of amino acids. This is often referred to as a code. But functionally, DNA is a template. The use of words like 'code' can be misleading. If we want to leave you a secret message, we might put it in a coded form, and if you had an Enigma machine, or its modern equivalent, you might be able to understand it. But this is not how life works. We use a series of letters (A, T, C, G) to represent the base pairs in the DNA. But that then is only a code because we have created it to represent the base pairs. Nature does not need a code. It already has the base pairs, and it does not need to represent them – certainly not as letters of an alphabet.

The Genome Is Not a Central Organiser

So DNA is not a secret, and certainly not the 'secret of life'. It remained a secret only in as much as our ingenuity could not immediately see it or understand its function. If there is an enigma in life, then it has been created by us as a ghost in the machine, controlling what organisms do. It might be appealing to see all the letters of the DNA sequences appear on a computer screen, but that is not how the cell sees it or uses it. No cellular site-foreman opens the 'blueprint' on a table from which he then builds the cell. The DNA base pairing is simply an essential chemical fact. Indeed, there are several essential amino acids that the body cannot manufacture, which must be obtained by organisms in their diet. Thus, we do not have genes for these amino acids, and nor do we have genes for a great deal else, including all the membranous processes where much of the complex control of cells occurs. Even the nuclear membrane is part of this active mechanism whereby the cell can select what has access to the DNA. So, if DNA is not a code, what exactly are people doing in genomics?

Genomics Is Probabilistic

In genomics labs, scientists will look for associations of function or disease with particular DNA profiles. But such links are rarely one-to-one. The greater the number of people in their samples, the greater the chance that they will find such associations. But consider this: if we need a million people to find such an association, how could our cells interpret such 'information' functionally? Cells could not translate DNA in such a probabilistic way. DNA is engaged in the cell's biochemistry rather than in probabilistic calculations. A particular DNA sequence might indicate a disease risk in the population or for a given individual with that profile. Still, it is one of many possible such outcomes. At best, if indeed it is a code, then it is a vague and variable one. It is we who are treating it as a code. Treating DNA as a code misrepresents how cells use genes. How a gene is used or expressed may vary with the functional context. Genes are multipurpose. Indeed, many geneticists favour an 'omnigenic theory', where all genes are involved in most, if not all, functions.

The Central Dogma of Molecular Biology

Where, then, does the omnigenic theory leave the Central Dogma? We have this beautiful process of transcription of DNA through RNA to protein. This leads to another error. A key proposition of the Central Dogma is that cells cannot alter DNA in a directed way. Furthermore, there is assumed to be no facility for such alteration. However, this proposition is not based on experimental evidence, and there is a growing body of evidence that it is false. It is based on the idea that the transcription process from DNA to protein cannot work backwards. But the cell does not need to do this to alter the genome. In one sense, it merely needs to allow changes to occur, as it can, for example, by varying the error correction process, or by large-scale chromosome and genome rearrangements, and then choose between the variants.

One-Way Flow?

However, the Central Dogma of molecular biology contends that this transcription cannot work backwards: a DNA sequence cannot be specified from a protein sequence. Thus, the causal chain of causation can only work one way, from DNA to protein. This is pivotal, because it would be difficult for the organism to influence the DNA or change it in any directed way if it holds. It is called dogma because it is held to be incontrovertibly true. This forms the basis for considering genes as primarily causal: genes drive the system, and not the other way around. Yet it stems from a misunderstanding of how cells regulate DNA expression and how they may also alter the DNA itself. Cells can influence the DNA, not through back-translation from protein to DNA, but by using the processes that enable the DNA to be corrected, and precisely this mechanism operates in the immune system. Thus, DNA is regulated by the system, and it can also be altered by the organism.

It is worth considering how the Central Dogma, for which there was little empirical evidence, could gain such a hold on scientific thinking. It was done by the stroke of a pen rather than real scholarship. It simply had to be true, for if it were not then our view of genes would be different. In his autobiography, *What Mad Pursuit*, Crick wrote:

> I called this idea the central dogma, for two reasons, I suspect. I had already used the obvious word hypothesis in the sequence hypothesis, and in addition I wanted to suggest that this new assumption was more central and more powerful. . . . As it turned out, the use of the word dogma caused almost more trouble than it was worth. Many years later Jacques Monod pointed out to me that I did not appear to understand the correct use of the word dogma, which is a belief *that cannot be doubted*. I did apprehend this in a vague sort of way but since I thought that *all* religious beliefs were without foundation, I used the word the way I myself thought about it, not as most of the world does, and simply applied it to a grand hypothesis that, however plausible, had little direct experimental support.

Crick's statement is shocking. It shows that many scientists ignore the rigour of philosophical analysis, and also scientific method. Yet it provided a false foundation for the gene-centric view of function and evolution.

Whatever thought one has about that method, testability via experimental support is a crucial ingredient of science. Simply holding onto ideas as dogma regardless leads to other errors of thinking. Indeed, imposing such a doctrine has the harmful effect of preventing theories and hypotheses that counter it. Those who have proposed an alternative perspective have been regarded as heretics. Yet what Crick said is true: there is 'little direct experimental support' for the Central Dogma, and the contrary evidence was there to be seen.

This dogma is where the resistance to understanding organisms as creative agents in their lives and in evolution has arisen. This is how science gave agency to genes but denied it to organisms. But even the Central Dogma could twist and bend in the wind.

There are many misunderstandings about what the Central Dogma states and how it should be interpreted. Crick had two formulations. The first and earliest, in 1958, was the simplest: DNA forms RNA, which forms protein, but not in reverse. However, in 1970, in response to the discovery that DNA can form from RNA, Crick modified his formulation of the Central Dogma. He wrote:

> The central dogma of molecular biology deals with the detailed residue-by-residue transfer of sequential information. It states that *such information* cannot be transferred back *from protein* to either protein or nucleic acid. (our emphasis)

Notice however that even this revised version of the Central Dogma (Figure 2.2) is incomplete, since it does not include any flow of information from the organism that controls *patterns* of gene expression, and which might *initiate* or *control* insertion of DNA into the genome either directly or following reverse transcription from RNA. Yet, by reverse transcription it becomes possible to transfer sequences, including whole domains corresponding to functional parts of proteins, from one part of the genome to another. We know that this has happened during evolution. The idea that the genome is isolated from any functional influences on the sequences is therefore simply incorrect.

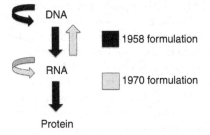

Figure 2.2   Central Dogma after discovery of reverse transcription. DNA codes for RNA, which then codes for proteins (black arrows). RNA can be reverse transcribed into DNA (upward grey arrow). The circular arrows represent the fact that DNA can also be involved in cut-and-paste modifications of the genome without the involvement of RNA, and that a similar self-templating can occur in RNA. This kind of diagram is often described as defining the information flows in biological systems. But it omits information flows that control gene expression and genome reorganisation.

Genome Reorganisation and Jumping Genes

Organisms can achieve genome reorganisation in several ways, not least by splicing and translocating sections of DNA. Indeed, this process has played a significant part in adaptive evolutionary change. It pulls apart the Central Dogma. The genome can be reorganised. Yet such was the hold of the Central Dogma that the evidential basis for this could not be published in standard scientific journals for many years. The evidential basis is now substantial, beginning notably with Barbara McClintock's work in the 1930s, 1940s and 1950s. McClintock showed genome reorganisation in corn (maize), where large domains are moved from one region of the genome to another in response to environmental stress, and even from one chromosome to another. Eventually, when it became clear that many others had found the same process in other organisms, McClintock received the Nobel Prize in 1983. She wrote:

> In the future attention undoubtedly will be centered on the genome, and with greater appreciation of its significance as a *highly sensitive organ of the cell*, monitoring genomic activities and correcting common errors, sensing the unusual and unexpected events, and *responding to them, often by restructuring the genome*. (our emphasis)

At the time of her discovery, she would not have known that the restructured components of the genome were DNA, nor the details of communicating a signal all the way to specific parts of the genome. We now know that signals are transmitted from cell membrane receptors via microfilaments to the relevant parts of the genome. We also know that it is the cell that controls access to the nucleus by controlling channels in the nuclear membrane. This is another example of how the cell instructs the DNA. Cellular organelles, such as the nucleus, mitochondria, sarcoplasmic reticulum, ribosomes and so on are factories of the cell, but it is the cell as a system that controls them.

In the perceived wisdom of evolutionary biology, there are three key features. The first is that small mutations (single nucleotides) in the DNA are occurring randomly. The second is that these mutations are passively selected in nature (natural selection), weeding out what does not work. The third is that there is no control or direction to these processes. However, the work of McClintock and others does not fit this standard story. Instead, a growing body of evidence shows that the organism can control these processes in response to environmental change. Moreover, there are as many as 11 enzymes involved in the control of replication. Nothing could be further from the description of DNA as self-replicating like a crystal.

Natural Genetic Engineering

James Shapiro demonstrated in 1992 that a natural process of genetic engineering occurs in bacteria in response to environmental change, extensively documented in his 2011 book *Evolution: A View from the 21st Century*. Furthermore, the process of somatic hypermutation in genomes has also been discovered in bacteria and other organisms under environmental stress. Somatic hypermutation occurs when the cell allows mutations (errors) to remain uncorrected in targeted sections of the DNA. We are more familiar with this in the immune system, but control of the error-correcting mechanism is more general. These processes are often targeted at specific regions of DNA, as they are in the immune system. It is not beyond possibility that this control can also enable splicing and translocation. A process involving 11 enzymes is coordinated with complex feedback signalling. The details of this remain to be worked out, but many of those working in the field believe that failure in this process may be a major cause of cancers.

In any event, the idea that evolution occurs through the gradual accumulation of small mutations creates a problem. It is difficult to see how this sort of gradual accumulation of small changes would itself lead to the creation of distinct species, *speciation*, which is a major feature of evolution. Such minor modifications would simply be absorbed in the population. By contrast, evolution shows explosive periods of new species creation such as occurred in the 'Cambrian explosion', some 500 million years ago. There is now evidence that speciation occurs when significant changes are happening rapidly in the genome.

Genome duplication involves cells having more than two sets of paired chromosomes, and this is particularly common in plants. Genome reorganisation during hybridisation (cross between distinct species) is also a mechanism of speciation. Furthermore, the first human genome, published in 2001, showed that entire functional domains of sequences for proteins have shifted location compared with other species. The recent development of CRISPR (cutting and splicing DNA) technology is based on natural processes in bacteria and archaea (also known as archaebacteria, an ancient group intermediate between the bacteria and eukaryotes). Cut-and-paste genetic engineering occurs in nature. We humans have copied it to create our own editing of genomes.

Given the disappointingly low associations emerging from genome-wide association research, a greater focus on such genetic reorganisation and its functional consequences would be more beneficial. Indeed, genome studies have provided a better understanding of speciation by tracking such genetic reorganisation.

What Is Real and What Is Illusion in the Central Dogma?

Francis Crick's Central Dogma is real when it is taken to refer to the chemical fact that DNA sequences are used by living cells to make amino acid sequences forming proteins. It is illusory when taken to exclude control of the genome and its reorganisation by living organisms. Furthermore, control is not simply chemistry. Our Figure 2.3 makes that clear by including the influences on DNA that make physiological control of the genome possible. Those influences include environmental and psychosocial factors.

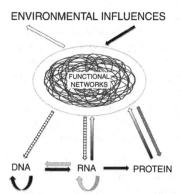

ENVIRONMENTAL INFLUENCES

FUNCTIONAL NETWORKS

DNA   RNA ➝ PROTEIN

Figure 2.3 Central Dogma embedded in living functional networks. This diagram adds to Figure 2.2 the information flows that control gene expression, i.e. transcription factors, methylation and interactions with histones. It also represents the use of the molecular mechanism of reverse transcription to reorganise the genome (natural genetic engineering), and the interactions with the environment.

The Child's Lego Experiment

One way to explain how genome reorganisation involving complete functional domains of DNA would speed up evolution is to imagine two children constructing architectural features from Lego bricks. You can effectively make any structure you wish from the flexibility of the Lego concept. The original small bricks can be connected to form any shape one wants, just as real building bricks can be used to construct a bridge, a small cottage, a railway signal box or a vast castle. Constructing with small bricks is laborious, but extremely flexible.

But, just as real builders today can use prefabricated structures to build much more quickly, it is also possible to buy ready-made architectural features with Lego-style connections so that they could easily fit into any other structure made using the original bricks. What would happen, therefore, if we gave two children some bricks and had a race? To one child we give just the original small bricks. To the other we give a mixture including ready-made arches. It is obvious that the second child will make a bridge much faster than the first.

Evolution has made use of this principle in rearranging genomes using cut-and -paste 'natural genetic engineering'.

Reshaping existing structures is how we get repurposing of structures and function, such as in the evolution of the middle-ear ossicles in mammals. Having three ossicles in the middle ear is a defining feature of mammals. Reptiles and birds have only one such ossicle, the stapes or columella. These little bones transfer vibrations. They evolved through novel functions of the jaw joint. The *Bapx1* gene is a jaw joint marker, expressed in the developing articular–quadrate joint in birds, fish and reptiles. In mammals, *Bapx1* is found associated with the malleus– incus joint in the developing middle ear. Evidence suggests that this reshaping has occurred through changing expression in embryonic development.

Is the Weismann Barrier Now Embodied in the Central Dogma?

Charles Darwin is famous for his theory of evolution by natural selection. Still, he also thought processes other than natural selection played a role in evolution, including acquired characteristics. That means that functions developed during the life of an organism may also be passed on to the next generation. Darwin even developed an idea of how that could happen. He was not alone: Jean-Baptiste Lamarck (1744–1829), who had also proposed it, sadly became the butt of many a joke in biology teaching. The idea of acquired characteristics was simply rejected and unexplored.

In his 1883 lecture, August Weismann (1834–1914) not only claimed that natural selection was sufficient for evolution, but sought to eliminate the inheritance of acquired characteristics from Darwin's theory. He did this by inventing a trick, the idea of what has become called the Weismann barrier. This is the theory that the germ cells, eggs and sperm, cannot be influenced by any changes in the organism acquired during its lifetime. This idea became embodied in the Central Dogma.

Many of the changes in the genome are part of the ongoing physiological function of organisms. The question then is whether and how any of these can be preserved and passed on to future generations through the germ line. It is often argued that they cannot because the germ line is protected from these changes by the Weismann barrier. There is no evidence for such a barrier. This idea of a barrier preventing change in the genome is often confused with the Central Dogma itself. They are however two distinct ideas concerning different levels: cells and molecules. Access to the germ line is controlled. An increasing body of evidence shows transfer of somatic changes (change in body function) across generations through the germ line. It is known, for example, that epigenetic changes (genetic modifications that impact gene activity without changing the DNA sequence) are communicated through the germ line, persisting for several generations.

Inheritance of Acquired Characteristics

This brings us to a crucial question. Can adaptation in one generation be passed on to subsequent generations through the germ line? The answer is yes. Early-life exposures and metabolic adaptations in adult parents have been shown to modulate epigenetic regulation in germ cells, thus providing a non-genetic molecular legacy influencing the health of subsequent generations. Furthermore, the evidence is now abundant. Your doctor will undoubtedly ask one key question during a health check – whether your family has a history of critical metabolic diseases such as diabetes. The reason is that such conditions tend to run in families. However, only 5–10 per cent of cases have a clear genetic origin. The 'missing heritability' is due to shared environmental causes, particularly in early life. Yet even this is not the entire story.

Multiple animal models have demonstrated that early-life exposures experienced by the parent can impact phenotypes in the offspring and subsequent generations, even without further environmental stressors. Even acquired behaviours can pass across multiple generations through non-genetic factors. Let us clarify: acquired characteristics can pass through the germ line. As a recent extensive review concluded:

> In summary, it is clear that the paternal lineage is responsible for more than just its genetically-encoded information. A variety of distinct epigenetic mechanisms, such as DNA methylation, histone modification and small RNAs, may collectively reflect prior and current environmentally-determined phenotypes in the father, thus providing a robust legacy of critical molecular information for his descendants which contribute to

paternally-mediated inheritance of phenotypes and initiation of a vicious cycle of disease risk across generations.

It is often argued that such epigenetic changes will not persist for more than a few generations, or not become incorporated into the genome. However, evidence from studies of populations living at high altitude contradicts this assumption, where what must have been initially a stress-induced epigenetic change is assimilated into the genome. So populations of humans have adapted to the persistent hypoxia of high altitude in several locations, and recent genome-wide studies have demonstrated the genes involved. In some populations, genetic signatures have been identified in metabolic pathways involved, such as the hypoxia-inducible factor (HIF) pathway, which orchestrates the transcriptional response to hypoxia. In Tibetans, such changes are found in several genes involved in these pathways, and this is an example of adaptation acquired in genes orchestrated by the organism. The precise genetic changes differ depending on the geographic region and the species involved. The same is seen in other animals, such as birds. The important finding is that, while very similar characteristics are produced in the organism, the changes at the genome level can be very different. In the case of birds, the authors of one article even include this fact in their title: 'Predictable convergence in hemoglobin function has unpredictable molecular underpinnings'.

The Gene Delusion

The gene-centred dogma derives from at least five erroneous assumptions: the concept of the gene as a precise self-replicator; the view of the organism as simply a vehicle to transfer genes with no agency; that natural selection is a passive process in which organisms have no active part; that changes in the organism cannot be transferred across generations; that DNA is a code or blueprint instructing the organism in its function and behaviour. None of these contentions is true.

3 What Evolves?

Standard evolutionary theory represents genes as the target of evolution. But organisms may functionally develop without alterations in their DNA, and they may also buffer changes in the DNA to retain function. It is organisms that are the agents in the process of evolution. Outside a living system, DNA is inactive, dead. Furthermore, many significant transitions in evolution have not depended on new DNA mutations. They arose from the fusion or hybridisation of organisms with existing but different DNA. All the molecular processes in a living system are constrained by its purpose. Viewed this way, genes are the most constrained elements in organisms. Evolution of different species has occurred through extraordinarily creative and varied processes that include cooperation and fusion of existing species and the exchange of DNA and organelles. It is much more like nature using preformed tried and tested functionality than through slow gradual mutation. Evolution can occur in leaps and bounds.

Darwin's 'Struggle for Life'

Living systems continuously adapt to environmental change as they develop, as they age, and as they make decisions and alter their behaviour. These changes are physiological, behavioural and cultural, and some of them are passed on to future generations. Offspring are never identical to their parents. Change is an essential aspect of life. Without it no organisms could survive, nor could they evolve. Reproduction is part of this process of adaptation and does not serve simply to replicate.

So, what evolves and how and why do evolutionary changes occur?

Charles Darwin could see how differences in individuals of a species could lead to different survival rates in what he called the 'struggle for life'. He called his 1859 book *The Origin of Species by Means of Natural Selection* because he could see that those best adapted to change would produce more progeny. The offspring may resemble their parents, but they are not identical. These differences are vital in the process of evolution because they are the sources of adaptation and selection. If reproduction did not produce change, or if organisms could not alter their functionality, then there could be no evolution by natural selection. Evolution thrives not on immutability but on change. This is why focusing on the idea of an immutable gene as the target of evolution is an error. If anything, evolution targets that which changes and, crucially, that which engages ecologically, which is the functionality and behaviour of organisms. In *The Origin of Species*, Darwin wrote:

> As many more individuals of each species are born than can possibly survive; and as, consequently, there is a frequently recurring struggle for existence, it follows that any being, *if it vary however slightly in any manner profitable to itself*, under the complex and sometimes varying conditions of life, will have a better chance of surviving, and thus be naturally selected. From the strong principle of inheritance, any selected variety will tend to propagate its new and modified form. (our emphasis)

That, essentially, is his theory of natural selection. Notice that he refers to the propagation of 'its new *and modified* form'. He knew that reproduction was not just replication. It is itself the origin of novelty.

What Is Natural Selection?

The first chapter of Darwin's book explains why he called the process *natural* selection. He did so in contrast to the *artificial* selection carried out by breeders of dogs, cats, cattle, sheep, horses, birds, fish, plants and many other species, where the selection is produced by the *active* role of the breeders choosing individuals with the characteristics they wish to preserve and develop. Natural selection by contrast is thought not to involve any choice by any individual. So differential survival in natural selection is regarded as more like a *passive* filter.

This is wrong, because the environment itself is in the largest part other active organisms. Over time it is organisms that are selecting. Evolution is an iterative ecological process. The selection process is behavioural and physiological.

Gene or Phenotype?

What, then, did Darwin think was the object of natural selection? Since it depends on the differential survival of individuals, one obvious answer is individuals. His work showed that species had changed with time as some individuals survive better than others, and so species could develop through selection of the phenotype. This is not the position of neo-Darwinism, which focuses on the gene as the target of evolution rather than the phenotype. So here is the major disagreement: gene or functionality. We might be tempted to say that both are correct, but focusing on functionality alters our perspective profoundly. Organisms are no longer passive vehicles, but active parties in the process of evolution.

It Takes Two to Tango

Giant pandas, *Ailuropoda melanoleuca*, have usually been regarded as solitary creatures, coming together only to mate; but recent studies have begun to reveal a secret social life for these enigmatic bears. GPS tracking shows they engage with each other more often than previously thought, and they spend time together. What we do not know is what they are doing when together. Perhaps it is an exchange of information, or a bonding with a potential future mate. And whether or not they are physically 'together', they may be constantly aware of each other. Whales, for example, can be hundreds of kilometres apart in an ocean, yet constantly communicating. Their choices and behaviour are vital to their survival.

For such large mammals, pandas have relatively small home ranges. This is not surprising. Pandas feed almost exclusively on bamboo. The only real threat to pandas is from humans. No wonder then that the panda is the symbol of the World Wide Fund for Nature (WWF). Pandas communicate with one another vocally and by scent marking. They spray urine, claw tree trunks and rub against objects to mark their paths, yet they do not appear to be territorial as individuals. Their behaviour informs others of their presence and their state.

Vegetarian Carnivores?

From an evolutionary and physiological viewpoint, pandas are an enigma. They are 99 per cent vegetarian, but their digestive system is more typical of a carnivore. For the 1 per cent of their diet that is not bamboo, pandas eat eggs, small animals and carrion, and they are known to forage in farmland for pumpkins, kidney beans, wheat and domestic pig food. It is thought these bears switched to eating bamboo, partly because of its abundance, and because they don't have to fight with other animals to get it. That was a significant behavioural change. Bamboo is high in fibre but has a low concentration of nutrients. So pandas must eat copious quantities of stuff every day.

The Social Life of the Panda

It is not clear if giant pandas are promiscuous, but they mate more successfully when they are free to choose their mate in the wild. This could also explain the poor breeding success in captivity. Mate selection is a significant ingredient in reproductive success across all species and is vital for species adaptation. It plays a crucial role in active selection in evolution. It is often not a momentary decision. When choosing a mate, a panda does not see genes; it sees the behaviour and function of another panda. When choosing a carpenter, we do not inspect their tools; we regard their proficiency, demeanour and work they have already done, their experience, how they behave.

Pandas sleep a lot, taking regular naps throughout the day simply lying on the ground or resting against a tree, balancing on a branch. Of course, much of their time is spent eating. But there is so much of a panda's life that we do not see that other pandas do see, sense and know. The pandas have a history of engagement. This is perhaps more apparent in socially tight species, where physical interaction is regularly ongoing. We humans are very much like that: our social being is the more significant part of our environment in the process of evolution. But all sexually reproducing organisms have interaction and a selective process.

What Controls Organ Size?

But we learn something more from the giant panda. The iconic black and white panda, the Sichuan giant panda, is one of two subspecies (Figure 3.1). Their cousins are the brown-furred Qinling giant pandas. Comparing their genomes shows that the two subspecies diverged around 10,000–12,000 years ago. For the size of their bodies, giant pandas have remarkably small organs, including small penises. In addition, several genes associated with reproductive performance, including sperm production and male genitals, have unique evolutionary traits in the giant panda genome, which may be partially responsible for the giant panda's low reproductive rate.

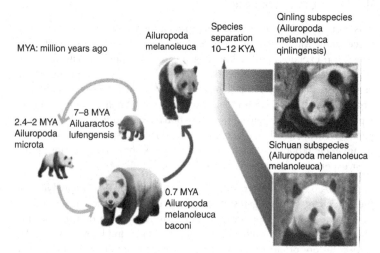

Figure 3.1 Evolution of the modern giant panda subspecies. The modern Qinling and Sichuan populations split 10,000–12,000 years ago, forming the current subspecies distribution.

Recent studies show that organ size is regulated at cellular and tissue levels by a cascade of regulatory proteins called the Hippo pathway, limiting the expression of genes related to cell proliferation. The system is constrained both by factors inherent in the cells and by the relationship with neighbouring cells in tissues, such as cell polarity and cell-to-cell adhesion. Growth is constrained by neighbouring cells. It is an example of how control of gene expression exists at an organisational level. In the evolution of the giant panda, genes involved in this regulation have changed or have been deleted.

These changes in the giant panda are relatively recent in evolutionary time. They are not gradual mutations of single genes accumulated over time; there are significant changes or deletions of genome sections. In this case, it appears that changes in the regulation of growth are a prime factor. But it takes two to tango. Indeed, those changes in the genome may confer some balance of advantage, but the novel traits must then be selected by other pandas.

Therefore, in order to understand the control of growth, development and evolution and what determines it, we must look at function at different levels: social, behavioural, organ system, tissue and cell. It does not come from the DNA alone.

Evolutionary Transitions – Hybridisations

Evolutionary transitions, or identifiable changes adopting novel functions and creating new species, can be caused in a variety of ways. Much depends on the level of organisation in integrative function. The earlier stages of evolution may simply involve the coming together of distinct types of single-celled organisms, a cooperation and eventual fusion between two prototypes of life. This symbiogenesis is almost certainly what happened in the evolution of how energy is mobilised in animals and plants. All life mobilises energy to do things, and to do this it needs a way to transform and store energy.

Our lives depend on the mobilisation of energy from a particular molecule, adenosine triphosphate or ATP for short, where energy is stored in the bonding of the three phosphates. This energy can be released by breaking the bonds. ATP is mobilised by specialised organelles, the mitochondria, the power packs of our cells. But these power packs did not always exist.

Mitochondria are the energy factories making ATP in animals and plants. But plants use another factory system enabling them to absorb light to create and store energy, the chloroplast. But even these organelles are found in some animal species. The panda, like all animals, must derive its energy from eating either plants or other animals. Even some plants do that. Early in evolution, the mixing and sharing of bits of machinery was common. But mitochondria do something else important in evolution: they carry some of the DNA. In origin, these are thought to be the circular genomes of bacteria engulfed by the early ancestors of today's eukaryotic cells. Over time, some of this DNA has been transferred to the DNA in the nucleus.

By any criteria, the formation of cells that can produce the energy molecule ATP more efficiently using mitochondria was one of the most significant steps in evolution. Being able to produce more energy led to enhancing the production of more proteins and many other molecules that can form the structure and provide the functions of organisms. In turn that meant that organisms could become larger. They could become multicellular. All the forms of life that we can recognise today without having to use microscopes evolved because of this transition.

Notice that, because it was the cooperative fusion of two *existing* organisms, it did not depend on slow accumulation of gene mutations. It consists in the, probably sudden, mixing of two independent genomes and in the fusion of the two cellular structures from the two organisms.

Furthermore, the subsequent reorganisation of the genomes in the fused organisms is typical of what organisms can do. Organisms can move genes around according to the needs of the organism. That happens when there is mixing of genomes and the process turns out to be a major driver of evolution. Changing the energy source enables other evolutionary pathways to follow. But this can also arise without the fusion of two species.

Cooperation is evident in ecosystems. For example, different species may form interdependencies in obtaining food, each fulfilling a necessary role. Such dependency, as with predator–prey relationships, will act as an iterative driver in evolution; change becomes purposeful in that nested function. The better adapted to the cooperation, the more successful it becomes. It is in this sense

. that we say that evolution is partially directed. Again, the cooperation between species can allow evolution to take different paths.

Mixing It Up with Hybridisation

Animal and plant breeders have known for tens of thousands of years that crossing varieties can produce new forms. For example, since the first domestications of dogs and sheep, many types have been made, creating and moulding them for a purpose. So, likewise, we share some of our genomes with the now extinct Neanderthals.

The evolution of new varieties can happen relatively fast. It happened in recent history when, around 30 years ago, a lone male finch flew from one of the distant Galapagos Islands to a more central one. It could not find any females of its own species, so it mated with one of the female finches common to that island. The result is a new species of finch, established within just two generations. This recent speciation has been studied carefully over the entire 30-year period. As with the symbiotic fusion between two microorganisms, such hybridisation can trigger speciation.

Reproduction is a way of mixing and sharing the genome. All sexual reproduction is brought about by the fusion of the sperm and egg with their different genomes. That is why we humans have two versions of each gene and other sequences in our genomes. Notice also that sexual reproduction automatically means that the progeny cannot be identical to the parents. As Darwin knew, they are not clones of one parent or the other. That is clearly true of human evolution. Consider how varied we are as individuals, even within a family, even when we can recognise similarities. This is why genomes can be used to track significant evolutionary change, forks in the path in the direction travelled. Therefore, we say that reproduction is not simply a matter of producing a copy; its purpose is not replication but the creation of novelty. This is important for organisms because they experience continual changes in their environment and their interactions with other organisms, including other species. Nothing is static. The very essence of the evolutionary process is the achievement of integrity through change. Just as viruses are continually mutating to get past our immune systems, we constantly change to deal with the challenges we face. Life creates problems, but it is also the means for developing solutions through

Major transitions in evolution

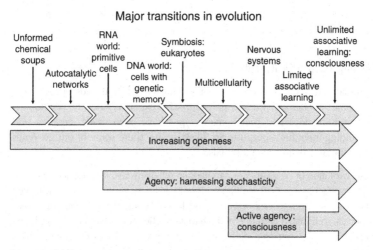

Figure 3.2   Evolution of organisms represented as a possible time sequence, showing major evolutionary transitions.

this continuous process of change. Life does not seek to maintain a particular bit of itself, the gene; on the contrary, it seeks to change it, or at least shake it up a bit.

Figure 3.2 distinguishes nine major stages in the evolutionary process leading to organisms like us. Each transition from one stage to another facilitates later transitions (the ratchet effect). Organisms become increasing open to their interactions with other organisms, leading to the forms of learning and anticipation involved in active agency. Harnessing stochasticity forms the basis for organisms editing their genomes. Chance therefore is not just passively experienced by organisms, it is actively managed.

Making It Up as It Goes Along

These examples of major steps in evolution do not depend on the slow accumulation of new random mutations, as assumed in the standard textbook theories of evolution. Quite the opposite is true. So, how many of the main transitions in evolution are dependent on just random mutations accumulating to produce a new species? In other words, how much of the evolutionary process has been dependent on the gene-centric process of everything changing at the level of genes? And how much has depended on the alternative recombinant process of the creation of new levels of organisation, which then constrain what the genes do?

Early Life's Secret? Life Gets Constrained

The answer to the question of how living systems first formed is still unknown. But what we do know is that fusion of these early forms appears to have played a significant role in the appearance of eukaryotes (organisms whose cells have a nucleus and other organelles). The four steps before eukaryotes are much more uncertain.

Each of the transitions shown in Figure 3.2 involved the development of a new level of constraint on the molecules of the original unformed 'soup'. In the unconstrained state, any self-maintaining reactions that happened to form would quickly disperse in the liquid environment. The first transition may therefore have been the occurrence of such reactions within the constraint of tiny fissures in the earth's rocks. This is speculative, but some kind of structural constraint would have been necessary before the formation of a cell membrane (Figure 3.3).

Cell Membranes

When you put phospholipid droplets (fatty molecules with a hydrophilic or 'water-liking' end) into some water, an interesting thing happens: they can form lipid vesicles. In a sense, these are the sacks of life producing a boundary between what is inside and outside. It requires no DNA; it needs only physical chemistry. Any early self-maintaining reactions caught within the vesicle would then be constrained by the membrane and not disperse. The membranes of living cells, made of phospholipids, are functional entities with channels that control substances coming in and out. The membrane generates the electrical potential difference between the inside and outside of the cell, which also determines other organelle responses: the control of motility, secretion, and engulfing particles, and even the control of gene expression processes. So what

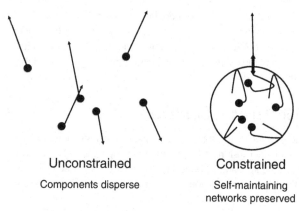

Unconstrained

Components disperse

Constrained

Self-maintaining
networks preserved

Figure 3.3   Unconstrained and constrained systems, illustrating the principle of constraint. Left: six particles randomly distributed in an unconstrained space. In this state the particles will progressively disperse. Right: The same particles enclosed within a controlled semipermeable boundary. In this state only particles that the boundary lets through can move beyond the boundary. In this diagram, for simplicity, we are assuming that all are constrained, except for one that happens to have encountered a channel in the boundary (the cell membrane).

came first? The chicken or the egg? In the early stages of life, the one occurs with the other, not before it. So, even in early life, there was no privileged level of causality. Life is an integrative self-conserving entity. In its initial stages it was likely to be very tentative and less structured; however it came about, phospholipids transformed its potential. And this is what evolution had to work with. Membranes were a necessary part of the origin of life.

It may have been this simple chemical fact about oil–water suspensions that got the very first compartmented structures going on the way to the evolution of cellular organisms. How this occurred is the real secret of life. Fatty acids, from which phospholipids are made, are created by life.

Self-Maintaining Networks

Such self-maintaining networks would have required the equivalent of catalysts, which are chemicals that speed up chemical reactions. Further speculation is that RNAs may have played that role. Unlike DNA, which is relatively inert as a molecule, RNAs can act as catalysts. From this view, DNA came later. It can only have done so when the self-maintaining reactions were sufficiently well developed to enable DNA copying. So, if true, DNA could not have been what conferred the characteristics of early life. In organisms today, DNA is never itself used directly to produce proteins. Instead, DNA sequences are always used first to make RNAs. Thus, RNAs also play a significant role in controlling DNA.

Despite the speculative nature of these transitions, they all have one feature in common: the formation of different levels of constraint on the self-maintaining networks. Any other speculative theory would have the same feature, since it would still need to explain how any early self-maintaining reactive networks could have survived rather than being dispersed.

These early transitions may have taken a very long time to become established. The further development towards the first 'vesicles' that could be called living organisms capable of cell division was a transition that imposed increasing constraints on the molecular level. It also allowed the evolution of a cell architecture for transport within the cell. Figure 3.2 shows only three transitions before symbiotic fusion (which would have occurred at the stage marked 'DNA world'). Yet there must have been many more stages leading to the development of the single cells without a nucleus or organelles and ancient forms distinct from bacteria. They are much more complex than the speculative early vesicular forms. The transitions to further complexity must also have brought further constraints. With increasing complexity in the organisation came both the need for and the processes of downward regulation. Life didn't start with a blueprint, and it never needed one, because it proceeded by trial and error. Life feels its way iteratively rather than following a 'Book of Life'. It did not have a set of instructions.

Onwards and Upwards after Symbiotic Fusion

We can then distinguish five further transitional steps: (4) the fusion referred to in the story leading to the first eukaryotic cells, (5) the transition to multicellular organisms, (6) the transition to organisms with nervous systems, (7) the transition to nervous systems capable of associative learning and (8) the transition to unlimited associative learning.

What's in a Nucleus?

Formation of eukaryotes. Evidence for the archaebacterial origin of mitochondria is solid. However, we suspect that other organelles, such as the nucleus (which harbours the genome) and the ribosomes (which use RNAs to construct proteins from their amino acids), may also have originated through further fusions. We do not speculate further here, except to point out that each stage creates yet another level of organisation and complexity of function. The genome within is controlled by the cell.

Better Together – Transitions to Multicellular Organisms

Transition to multicellular organisms. Unicellular organisms are not always isolated individuals. Even bacteria, algae and fungi show cooperative behaviour in many ways. For example, bacterial films formed of thousands or even millions of individual bacteria have chemical signalling processes for communicating and altering their behaviour as a group. Science has given it a name: *quorum sensing*.

Quorum sensing synchronises the activity of the individual organisms. Perhaps the earliest multicellular organisms were loose groups of collaborative single-celled microorganisms, acting together in specific ways. What they could achieve together was more significant than if they were apart. In any event, cooperation is an essential factor in all ecosystems. But there is another advantage of coming together as a group: the development of specialisation. Tighter associations may arise when the specialisation is so great that some give up or lose the ability to do things others can do. This is undoubtedly what we see in complex multicellular organisms. But not all multicellular organisms would have evolved in this way. Some may have developed when single-celled organisms reproducing by division simply failed to separate.

Higher-Level Constraint

Regardless of how it evolved, cell interaction is an example of higher-level constraint (influence) on lower-level processes. It is that constraint that determines how the genome in each cell is expressed, and how it might evolve. What then happens is one of the most significant examples of higher-level coordination acting as a constraint on lower-level operations. As a result, cells could become specialised, just as, in human societies, workers became expert in their respective skills. Similarly, cells could take on specialised functions such as feeding, movement, sensing or defence.

In humans, for example, we can distinguish at least 200 different kinds of cells in our bodies. They are as diverse as bone cells using calcium to form rigid structures and muscle cells using the same calcium atoms (in their charged ionic form) to signal rapid movement. The coordination of all the different kinds of cells and tissues is a property of the body's tissues, organs and systems. All normal cells are subject to these constraints. When cells go 'rogue' and act against those constraints, we have what we call cancers. We are alive only because those constraints usually work well.

A Microscopic Key to Evolution – *Volvox*

Examining pond water with a microscope in 1674, Anton van Leeuwenhoek observed many single-celled organisms, which he called 'animalcules'. Some formed colonies or clusters, others were clearly multicellular, and one of these was a tiny green alga now called *Volvox*. Could this tiny organism be a microscopic key unlocking evolution's mystery?

Volvox is a spherical multicellular green alga, containing many small biflagellate body cells and a few large, non-motile reproductive cells called *gonidia*. As a result, swimming with a characteristic rolling motion, its name is 'the fierce roller' or *Volvox*.

Imagine the thrill. None of these microscopic animals could be seen with the naked eye. The microscope revealed another living world. This was like landing on another planet. Microscopic life forms, changing our view of life on earth. For such observations led to a fundamental theory of biology – that living organisms are made up of cells. But it also provides a key to understanding microscopic evolution.

Whilst Anton van Leeuwenhoek watched one of these creatures, a *Volvox*, he witnessed for the first time something remarkable – a microscopic birth. Of course, this is a key ingredient of microscopic evolution, carrying change to new generations. Noting that each of the small creatures 'had enclosed within it 5, 6, 7, nay, some even 12, very little round globules, in structure like to the body itself wherein they were contained,' he wrote:

> While I was keeping watch, for a good time, on one of the biggest round bodies ... I noticed that in its outermost part an opening appeared, out of which one of the inclosed round globules, having a fine green colour, dropt out, and took on the same motion in the water as the body out of which it came ... soon after a second globule, and presently a third, dropt out of it; and so one after another till they were all out, and each took on its proper motion.

A keen observer would see that some of these single-celled organisms form clusters. The cells in the cluster or colony look alike. So, easy enough then to envisage that more complex multicellular organisms might have evolved from such colonies. Living together may enhance nutrition and afford protection – another key to microscopic evolution.

Cell Specialisation

But living in a cluster also does something else. It allows cells to become functionally specialised, some becoming cells for movement, others for obtaining nutrients, and others for reproduction and dispersal – a mutually beneficial arrangement and a key ingredient for microscopic evolution.

The push and pull on the microscopic evolution of cells within the colony will then be the cooperative of cells in the colony. Single-celled organisms are multipurpose, doing everything to maintain their existence – moving, feeding, reproducing. But, freed from the need to obtain its own food, a unicellular organism in a colony can hone its specialisation, say, for movement or reproduction.

Gene mapping shows that *Volvox* evolved from loose colonies of single-celled organisms at least 200 million years ago. The cells of *Volvox* were now joined in a single evolutionary fate, a new landscape for adaptability. It is a transition that occurred hundreds of times to create different multicellular species, each with its own complex life cycle and characteristics. What this meant was that one for all, and all for one, became the standard for fitness. After all, cells could be sacrificed for survival of the organism.

Trial and Error – Spinning the Wheel of Fortune

Colonial life allows something else important in microscopic evolution – trial and error. Life can be a bit like a fruit machine, where you need three lemons, say, and you have two. Holding the two lemons allows the wheel to spin to find the third. Three lemons are better, but two will do for now. Evolution is like a continually changing jigsaw puzzle, where some parts fit better than others. Life continuously invents new ways of doing things in an ever-changing tapestry. Colonial living protects the organism even when spinning the wheel might produce results that are potentially harmful.

In this sense, life spins the wheel of invention all the time. It is never quite happy with things the way they are. Life is a bit of a gamble, but life can stack the odds in its favour – at least for a time. This is good, because change is all around us. Life does not really have the option to stand still, although some organisms may become dormant or have a dormant phase until favourable conditions arise. Seed dormancy, for example, is an evolutionary adaptation that prevents seeds from germinating during unsuitable ecological conditions.

Nervous Systems

Transition to nervous systems. In the early multicellular organisms, the con-straints, or functional coordination, depended on chemical communication from one part to another: hormones and transmitters. But chemical diffusion is a relatively slow process. Action often needs to be faster. If a squid needs to escape from a predator very quickly, it does so by sending a fast electrical signal to trigger a form of jet propulsion. It has a specialised bit of its nervous system to do this: the giant axon. It is an example of how functionality drives evolution. The conduction speed of the electrical impulse along the axon is dependent on its diameter. The larger the diameter, the faster the speed of transmission. To achieve this, the axon has evolved and developed from the fusion of many nerve cells to produce the giant one. At one moment, the predator has the tasty squid almost in its mouth; the next moment, all the predator tastes is some nasty black 'ink' while the squid has darted off to somewhere else. Nervous electrical activity is yet another example of a regulation that involves not a single gene. The protein molecules that can conduct the charged ions, like calcium, sodium and potassium, are all coordinated by the same electrical process or voltage that they generate. Another way to increase speed of signal transmission is to insulate axons. This also involves a cooperative process of cells working together. Some cells (Schwann cells) wrap themselves around the axons forming an insulating sheath; the sheath along the axon has gaps (or nodes) and the signal is conducted rapidly by jumping from node to node. Such signal transmission is involved in the control of the fingers in writing the words on this page.

The squid has even more evolutionary surprises up its sleeves (tentacles): it has over 500 million nerve cells (neurones) in its brain, consisting of two paired cerebral ganglia, more than twice the number in a rat's brain and as many as a dog. Two long tentacles are used to grab prey, and the eight arms to hold and control it. Science has long been fascinated by the extent of squid intelligence and problem-solving. On either side of the head, the paired eyes are housed in capsules fused to the cranium. They have an uncanny resemblance to the eyes of a fish, an example of convergence in evolution: the independent evolution of similar features in species of different periods or epochs in time. If you think that is surprising, then consider this. There are around 86 billion neurones in our human brains. And that leads us to another exciting feature of evolution: consciousness.

4 Purpose in Life

We are writing this book as agents with a purpose. Agency and purposeful action is a defining property of all living systems. Yet modern science has presented a reductionist, gene-centred view of life, where life is reduced to biochemistry, particularly DNA and proteins. It has even carved out its own areas of study – genomics and proteomics – as if these components can be understood in isolation from the organisms themselves. But they cannot. The gene-centric view of life creates a fundamental problem. If all action can be reduced to genes, or is controlled by them, then purposeful agency cannot exist. Indeed, it has been referred to as an illusion. At best, modern science gives this problem to philosophers, assuming that the answer does not lie in biology itself. This is a mistake. Casting the issue aside ignores the most creative aspect of living things: problem-solving and the agency of organisms.

Not in Our Genes

Modern biology has tended to present a gene-centric view of life, where genetics, or DNA, is thought to be a primary causal factor in function and behaviour. It is embedded in our language, where DNA is often referred to as 'a set of instructions', a 'blueprint' or 'a code' – so much so that organisms have been considered mere 'vehicles' transmitting genes from one generation to another. To this end, it is suggested the transmission of genes is the ultimate cause and purpose of the behaviour and function of organisms. In this gene-centric view, genes drive us and our decisions to preserve genes in a 'gene pool'. As we will present in this book, such a view gave rise to a fundamentally distorted view of nature, the idea of 'selfish genes'. In this book we show why this gene-centric view of organisms is wrong and based on false assumptions about how genes and organisms work.

The idea that DNA or the genome is somehow directing function, such that we can assign characteristics or types of behaviour to particular genes, is profoundly misleading. Given the complexity of organisation required for the function of organisms, it is impossible for this to be so.

Multicellular organisms can have many cells, the tiny building blocks of life. In the human body there are trillions of them, and hundreds of diverse types, such as muscle cells, skin cells, liver cells, renal cells, all serving distinct functions. Even with muscle cells, diverse types are arranged in particular ways to facilitate contraction and movement. Cells are specialised individually and strategically in the body. Furthermore, the genome does not exist as a distinct entity, separated from the cells, or in a strategic part of the body. It is in all cells and can only function as part of the processes within them. Metaphorically, it does not conduct or orchestrate the body in its moment-by-moment function. As we decide to do something, such as write the words on this page, our genomes are not choosing the words, or thinking the thoughts they represent, nor do they create our view of the page or the world. They simply are not equipped to do so.

The cells of our body are functionally organised into tissue types, and these tissue types are strategically arranged into organs and organ systems (Figure 4.1). For example, the circulation carries oxygen and removes waste products of metabolism, and also carries hormones or chemical signals produced by other organs, some specialised for so doing – the endocrine system. It is an example of how systems in our body form a meshwork of function. Our hearts, for example, pump blood around the body, but from moment to moment this is influenced by our emotions and other factors in this network. As described in Chapter 1, whilst we can find genes that are involved in generating the beating of the heart (the pacemaker and force of contraction), those genes do not control our emotions, and nor do they regulate the rhythm of our hearts. If we were to produce a diagram representing all the factors at work in the rhythm of the heart, it would be a very messy one, and it would be difficult to pinpoint a primary cause. Our bodies are awash with chemicals, but the production and distribution of those chemicals is not coordinated by

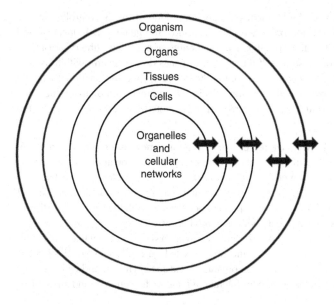

Figure 4.1   Living organisms consist of embedded levels of organisation. Cells contain the networks that carry out basic biochemical processes, assisted by the structures we call organelles ('little organs'). Cells are also the lowest level of organisation that can be said to be alive. In multicellular organisms, they are organised into tissues and organs, which together form the organism as a whole.

the genome, even though the genome is involved in producing some of them, specifically RNAs and proteins.

Our complex nervous systems are distributed throughout our bodies, branching through the tissues in intimate contact with the other cells. Also, our brains consist of billions of specialised cells, neurones, organised in complex ways to receive and send signals and release transmitter chemicals from trillions of synapses (tiny connections between neurones); and these are organised in even more complex ways on the processes of other neurones or on other cells, such as cells specialised in secreting hormones. Our skin is a complex sensory system which also provides protection and is a key component of temperature regulation. In the moment of happening, none of this is controlled by the genome. Yet, in contrast, the expression of the genome is influenced and controlled by this complex organised entity of our bodies (the organism). In this, the organism is self-directing. Our genes do not 'swarm within us' and control us; they cannot do so. On the contrary, they are functionally tightly regulated by the cells, and the cells in turn are regulated by the tissues they form, the tissues by the organs and organ systems, and those in turn by the organism.

In Figure 4.1 we have represented this arrangement as a series of concentric circles, but the influence of one level of organisation on another occurs in all directions (indicated by the double arrows). The circles are not fixed, or passive boundaries, they are functional entities; so, for example, the cell membrane is in an intimate process of exchange with the surrounding tissue, and cells communicate with one another, as do tissues and organ systems. This is what we refer to in this book as an open system. It is not like a machine, with cogs and wheels contained in boxes. In Figure 4.3 we extend this further, adding interactions between organisms and social groups, and with the eco-system. It is in these complex interactions that we find purpose and intention in function.

What Is Purpose – How Does It Arise?

Using a stone to crack a nut, a chimpanzee, *Pan troglodytes*, brings it down hard (Figure 4.2). The nut cracks easily, and the chimp gets a reward, the tasty morsel inside. Using the stone in this way is a skill this chimpanzee has only recently learned. His parents never cracked nuts like this. Nevertheless, the chimpanzee has used this stone before. He chose it as the best stone for the job, carefully assessing its efficacy by comparing it with other rocks. Some he discarded for being too small or too large, another was the wrong shape, but the one he now uses he had selected – and he has gradually shaped it to his needs. It is his best stone, and he keeps it in a safe place for future use.

This chimpanzee's success at cracking nuts with a stone has not gone unnoticed. Other chimpanzees have been watching. They sit in a circle observing; some are already selecting stones themselves and are using them to crack nuts. However, it is the first time they have done so. Learning from the

Figure 4.2 A chimpanzee cracking nuts.

other chimpanzee's example, they are also searching for 'good' stones better suited to the task. Learning from each other, some are better at it than others. Some are more experimental, a key ingredient of creativity.

Not so long ago, none of the chimpanzees used a stone. Now, most of them do. It spreads rapidly through the group because it is a creative solution to a long-standing problem: how best to crack nuts that are harder to break. Cracking nuts is intentional – it serves a purpose, to feed oneself. It is also an example of intelligence, using experience, knowledge and logic to solve a problem.

Just as we talk of stone-age humans, so we should recognise stone-age chimpanzees. The use of tools demonstrates the use of knowledge and creativity in solving problems and doing things in diverse ways. Chimpanzees develop techniques – in that sense they are technological, for example using a stick to get termites from a nest. But this use of tools shows another fundamental of life: agency. The chimpanzee acts with intentions; with purpose, the chimpanzee reasons. Achieving goals with tools is a clear example of intentional behaviour – it is also an example of logic and motivation that is contextually driven. A stick or stone may be shaped to better fulfil a purpose. That shaping is intentional. This learning is not in the genes, although it might change the way they are expressed in different circumstances.

Intention, Reason and Logic

When chimpanzees choose a stone or make other choices, they are doing something particularly creative in the universe – they are thinking. They are using reason and logic to make choices. This is intelligence, and organisms use intelligence in developing and using tools. They also use intelligence in anticipating the behaviour of other organisms, and often in working with them. The ecosystem is full of such interactions. Indeed, for some problems, solutions emerge through the evolution of ecosystems, which is driven by organisms. Organisms build or modify their niche (the immediate environmental conditions in which they live, and the interactions with it and with other species), and it is in this sense that the niche has purpose. Organisms are not separate from their environment but an integral part of its functionality. It is only through this that we can understand their behaviour, logic and purpose. So organisms are adapted to their habitat, but part of that adaptation consists in their ability to make it more suited to their needs: building nests or shelters, storing food, marking their territory, informing others of their presence. Just as a chimpanzee may hone a stone or stick, so organisms hone their immediate environment, actively adapting to change in a functionally purposive way. In this, we see purpose in the situational logic. For example, birds building a nest to lay eggs and rear their offspring, or beavers building a lodge for protection from predators and rearing their young. All this activity also changes the immediate environment and thus the potential habitats for other species to occupy. In humans we see this in our built environment and the elaborate social infrastructure that supports our needs and activities, but we also see it in our psychosocial interactions, and these in turn influence how we perceive ourselves, the world, and others.

Such interactions between organisms in the creation of their niche is represented in Figure 4.3, where the organism is presented as the primary agent of change through several levels or layers of organisation, represented by concentric circles in the diagram. At the core, there are the actions made by individual organisms on their immediate environment, or their reactions to it; then there are the interrelationships between organisms, either of the same species or of distinct species; next are actions taken as groups or as part of a society of organisms; but there are also interactions between social groups, and with the ecosystem. In this complex interplay, cause and effect run in all directions. Once again, it is an open system, so that what happens at one level

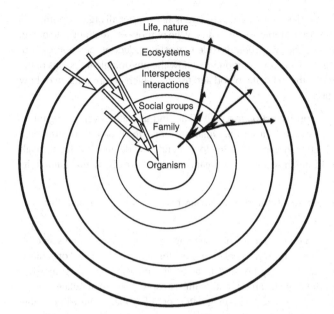

Figure 4.3   Diagram representing the interrelationships in the agency of organisms. Organisms are influenced by their relationships with others of the same species and other species, their role in social groups, and the interactions between social groups. In this book, we refer to these interrelationships as life or nature.

can influence what happens at another level. This is what we call nature, or simply 'life', in this book. In this process, far from actions at any level being determined by genes, what happens at the outer levels of organisation can influence the function of genes. Mutual grooming, for example in a troop of monkeys, alters hormonal expression and behaviour, enhancing harmony and cooperation. Furthermore, these complex interrelationships have evolved and are evolving in a dynamic way and have directionality (purpose) in that context.

Thus, the arrangement of muscles around the skeleton of the arms and legs of a chimpanzee allows it to manipulate a stick to get termites or a stone to crack nuts, and so much more. Not only is the external object used for a purpose, so too are the muscles, skeleton and arms, as is also true for all organ systems. It is in this sense, for example, that we can say the purpose of breathing is to obtain oxygen and get rid of carbon dioxide (a waste product of metabolism), but breathing is also a major component of how the physiological systems of the body maintain the pH of the blood and tissues, which in turn influences how cells function.

So we see purposive action (agency) all around us and in every living thing, because it is what organisms do. Organisms solve the problems of sustaining their existence by doing something to that end. For, definitively, life through the action of organisms is a self-organising entity creatively maintaining its integrity.

When we talk about purpose, we mean two different but related things in functionality:

1   What function is served, what the thing does – for example, muscle cells contract, creating movement or tension; this is a capacity to act or achieve something and may be used in achieving diverse kinds of goals. The arrangement of our fingers and thumbs enables us to manipulate objects in many ways, including typing letters on a keyboard. We will call this 'facultative purpose' in this book.
2   Purpose in relation to achieving goals – for example, the use of movement or action directed to specific goals or outcomes. We will call this 'goal-directed purpose' in this book.

In the context of the second of these meanings, we may ascribe intentions. We may understand this kind of purpose through situational logic. An organism acts to achieve a given outcome or outcomes, such as cracking a nut.

The distinction is important. The first can apply to a tool, such as a stone, say, or a stick for which an organism creates a purpose. The purpose of a nest constructed by birds is to protect and rear their offspring. The second applies where the organism intends that purpose. Much of life's intelligence is to do with this second meaning, about which we often use the question, 'why did he, she or it do that?' rather than simply 'why did it happen?' If we see a row of dominoes falling, we might explain it by physics and geometry alone. But the first domino to fall might have been pushed, and why it was pushed has a different kind of answer. It is to do with what we call motivation or intention. It is often hard to answer, yet it plays a significant role in biology. In many respects it is part of life making it up as it goes along. It is in-the-moment creative and may involve reason and logic. This leads us to another distinction.

Science has also tended to muddle two different issues: (1) why an event or behaviour happens, and (2) why a type of behaviour persists or happens commonly in a population. The first has an answer in the immediate sense; the answer to the second is in long-term outcomes for the population or species, such as fitness to survive or reproduce. The survival of an individual organism may depend on the former; that of the species may depend on the latter. So, for example, understanding why Jack is playing golf at a given time may involve many reasons, including simply that he enjoys it. This type of behaviour may have advantages, such as keeping fit or improving prowess, concerning fitness and survival. So we can distinguish the immediate cause from an ultimate or longer-term outcome, which the organism may or may not know. For example, we might find that playing games, by enhancing fitness, 'maintains genes in a gene pool', but it is not why Jack plays golf. Jack plays golf regardless of any knowledge he has about genetics. Culture rather than genes plays a major part in the kind of games we play; the English play more cricket than Americans.

So what about evolution? Does purpose play a role? Science tends to avoid the idea of any direction in evolution, seeing it as a passive process of blind chance. Yet evolution involves changes in form and function, and these have a purpose, just as honing a stone makes it better suited to the task of cracking nuts, or a stick to extract termites, so changing the structure or function of, say, muscle, may improve its performance. And how well these changes improve performance will influence survival and thus give directionality. As we will set out in this book, both the processes of change of form and function and of natural selection are active and confer a direction on evolution. Metaphorically speaking, the 'watchmaker' may be blind, but she feels her way in the process of change.

The process of evolution is iterative, involving active change in form, function and selection. This does not mean it 'knows' where it is going. It simply means it is dynamic and in part physiological. In this, the goal is not to maintain genes in a gene pool, but to bring about change to enhance fitness. So, far from the gene-centred view that life is a vehicle for maintaining them, genes are changed, often in substantive ways, to enhance survival. This idea is also represented in Figures 4.1 and 4.3, where each layer of organisation influences or constrains the changes in other layers. Fitness in cellular function is influenced by neighbouring cells and tissues, and their role in turn is influenced by other tissues and organ systems, just as the organism is influenced by interaction with others and the ecosystem. How well functionality works within this interplay is what we mean by fitness. The key question then is whether organisms can respond across generations to change in any targeted manner. This book will explain how this is possible.

Learning and culture matter. What the chimpanzee does when making choices is not simply explained by an algorithm (a simple set of rules). It is an adaptable process only constrained by their capacity to act or, as Ginsburg and Jablonka say in their book on the evolution of consciousness, they show unlimited associative learning. Their ability can change with practice and by observing others. This is true when birds build nests, and even when they develop their song with a local dialect. The extent to which organisms can adapt in this way is typical of life and only constrained by their capacity. Therefore, genes cannot determine any given behaviour, even though they might both enable and limit the ability to act. Genes are tools, not masters. We humans cannot fly because we do not have wings, but we can make wings and fly, as in making an aircraft. As much as our genes are involved in the function of our muscles, they do not determine the direction of travel. This behaviour is not in our genes, and nor can genes determine the nature of our behaviour. For example, there are no genes specifically for designing an aircraft, or for deciding to be an elite athlete or a musician. Similarly, they make us neither selfish nor altruistic. Genes have no moral compass. In assessing and selecting a stone, the chimpanzee uses the capacity to choose, and in the moment adapts to the circumstances.

Solving Problems – Constantly Changing the Rules

Problems are solved by organisms, not by genes. There is no gene or genes for solving a jigsaw puzzle. There is contextual logic and choices of pieces that fit.

So fitness (the ability of a species to survive in the particular habitat over generations) is an iterative process requiring adaptability to change; that is, it is a repetitive process in a sequence of outcomes. It is as though the jigsaw puzzle changes each time the pieces are moved. Imagine doing a jigsaw puzzle if the picture continually changes. That is what nature (the interaction of organisms) does. For example, we will see in Chapter 6 how changes in population influence behaviour in wood mice. How does one play, say, chess if the rules themselves are changing? For that is what organisms encounter in their continuously developing challenges. If a new species 'invades' a fresh territory, it can evoke profound changes in resident species, such as when grey squirrels, *Sciurus carolinensis*, from North America have invaded the territories of European red squirrels, *Sciurus vulgaris*. The impact on red squirrel breeding success also impacts other organisms, and not least the trees. In evolutionary time there may be sufficient niche differentiation to enhance fitness. A great deal of this adaptability is physiological or behavioural in the immediate sense of the organism adapting to changing conditions; but it also may occur across generations where structural and functional change better adapts the fitness of the species to survive. A key question then is whether and how adaptation in one generation could be passed on to subsequent generations. Here we find a fundamental disagreement in modern biology between those who insist that characteristics acquired in one generation cannot pass to subsequent generations, and those, including the authors of this book, who see such transmission playing a major part in evolution.

Genes Do Not Make Choices

A chimpanzee has a problem, to crack a nut. The solution, selecting the right stone, is considered before, during and after the choice of stone is made. It is an iterative, creative process involving what we call situational logic and trial and error. The option is changed by the experience of making choices. Our genes do not make choices, yet our choices may alter the expression of our genes. So here are the key ingredients to this behaviour: a problem and a solution involving options and capacities to act. The action is goal-directed, and the goal provides part of the answer to the question. For there is often a reason behind the why; that is, the reason for a particular choice is also determined by choices already made, or other motivations and experience; such a choice is also open to revision, even in the act itself. For example, turning left at the road junction, Jack searched for Jill, but he also believed Jill turned left. Life has interweaving, open narratives with ever-changing plot lines. No wonder science finds intentions difficult. There is an interesting cultural heritage that has led to some countries, such as the United Kingdom, driving on the left, whilst so many chose to drive on the right. That is not in our genes; it is embedded in our history of conventions, dating from when people travelled by horse and carriage.

We have moods and might get angry. We argue, and we discuss. Often, we seek agreement. We act in ways that are not readily understood solely in the context of what is happening, because how we feel has history in our relation-ships. This history influences the way we think about each other and about events that are happening to us. It is part of the contextual logic of behaviour. Not all our behaviour is reasonable or follows a set logic.

We might seek to understand 'moods' simply by measuring the release of chemicals in the brain that cause the brain cells to function in a particular way. Thus, love, hate and desire might also be so explained. No doubt, such chemical processes occur when we are doing things or when we have the disposition to do something. But there is a question of what it is that comes first in this process. Moreover, these processes are continuous and reiterative. Thus, moods and dispositions can as readily alter brain chemistry as be caused by it. This is called *conditioned arising*, or a change in the state of functionality. Because we are not closed systems, there is an intimate relationship between the individual, others and their surroundings, and with themselves, creating what we call 'moods' or 'states'. It is as crucial in understanding ourselves as humans as it is in understanding the behaviour of other forms of life.

Agency

Wherever we look, we see life harnessing and doing things in processes that solve problems. Without life, there would be no problems. The issues are the creation of life itself, and so also are the solutions. Life creates both the problems and the solutions. This is what gives agency to the processes of life. A rock tumbling down a hill may have consequences, but it has no agency. Agency is a feature of life. Living forms have goals, objectives, such as building a nest, obtaining food, moving from A to B, finding a mate, and many other objectives, including sometimes having fun. Maintaining integrity gives rise to all issues and to all problem-solving, and in this sense, intelligence is a necessary capacity of life. No purpose is served in assigning agency to genes whilst treating organisms as playing no part in their being. To do so simply ignores physiology.

This idea of agency of organisms may seem obvious, yet, in general, science has not handled the notion of intentions or purpose very well. It usually seeks to avoid purpose, as if it is incidental to what is really happening. As a result, explanations become mechanistic, explaining *how* something happens rather than *why* it happens. Moreover, it avoids what it calls teleological explanation, which is another way of defining purpose. Indeed, teleology can be regarded as sinful in science, and students are taught to avoid it. This arises from an overly mechanistic view of life.

So how can a living organism constructed from the material of the universe have purposes? How can anything be *intended*? Philosophers and scientists have struggled with this problem for centuries. But it is the wrong question. Intention does not come *from* anywhere. It is definitively what living things (organisms) do. Furthermore, ideas generated at a cultural level (Figure 4.3) influence the purposeful state of living organisms.

This book aims for nothing less than discovering how intentions can be compatible with the fact that living systems are made from the material of the universe. The answer is counterintuitive: that purpose is created by the fabric of the universe as living entities (organisms).

Anticipatory Behaviour

Teleology explains phenomena, or happenings in nature, in terms of their purpose, *telos*, rather than the mechanistic causes by which they arise. Thus, we can explain how the planets orbit the sun directly from physics; we do not need to invoke a purpose – inanimate physics alone will suffice. So the chimpanzee's behaviour gives us a problem: it involves attributing purposefulness to his actions. Furthermore, this purpose is creative. It is the organism, the living being, that creates the motivation and sense for its actions. When we see a chimpanzee with a stone, we might *anticipate* that he will use it to crack nuts. But he might not. He might use it instead to break heads, or simply throw it, or put it to some other purpose that we have not yet seen. For purpose itself is creative, and this uncertainty is difficult for science. It is not so easy to predict. The reason for an action may differ even as the behaviour is physically the same. It is not easy to create mathematical models to describe it, particularly if it requires a dynamic in mathematics that would allow for the unexpected. Nevertheless, ignoring this creativity produces a poor understanding of chimpanzee behaviour, or that of any organism, and certainly if we did not attribute *meaning* to it.

Whether correct or not, the intentions we attribute to others become part of our psychosocial being and influence us just as much as, if not more than, the physical environment alone. For example, the understanding of intention to anticipate the behaviour of others, and in doing so, it may create anxieties, expectations, hopes and fears. And we may well use it in our own logic and decisions. This is what we call psychology. It is a considerable influence on our own sense of wellbeing and our behaviour.

So why then does purpose or agency give modern biology a problem? The answer can be found in a fundamental misunderstanding of the role of genes in living organisms. As discussed in previous chapters, a prevalent view in modern science, often called the Modern Synthesis, would have us believe that genes are the answer to everything, and that our entire existence is devoted to their preservation. Anything else in our behaviour, thoughts, beliefs, hopes and wishes is superfluous, illusory, or driven by genes. In Chapter 1 we showed how this view gained ground, and why it has no firm scientific foundation.

Open Systems

We may not always know why an organism behaves as it does, but this alone is not sufficient to deny its purposiveness. The purposiveness of others forms a labile ingredient of our social being, whether or not the other understands the rules of any game we may be playing. This is the nature of an open system. It cannot be closed by rules and cannot be simply algorithmic.

There is no reason for evolution to follow rules at all. Evolution is also an open creative process. Yet many scientists avoid attributing any directionality to it or agency in it. In that view it simply becomes a matter of random events over which life has no control. It is as if the environment merely becomes a static sieve, or filter, for fitness rather than an active, iterative change process. Even Darwin would find this strange, because he understood the way organisms may play a part in the creative process of evolution. As we were taught in school, Lamarck is often mocked for his concept of *acquired characteristics*. Still, it is one with which Darwin concurred and for which there has been and is a growing body of evidence.

The Problem of Change

Of all the characteristics of life, the most useful is the ability to mould its function to a continuously changing environment. But neither the organism nor the environment is passive. On the contrary, the changes in one will require changes in the other. Sometimes these changes are immediate and involve the physiological and behavioural faculty of organisms. Other changes may be longer-term; they may be social and cultural, but they may also include changes in form and function in the evolution process. Evolution is a dynamic process; fitness is continually honed. Imagine a lock for which you must find a key while it is changing. Nature is doing this all the time, as in the immune system, for example, as it continually meets new lock-and-key challenges when new viruses, bacteria and other antigens invade the body.

The Story of the *Euglena* Eye

Evolution has occurred in many ways. To adapt the well-known proverb, nature can be said to be the mother of invention, and this we see in the microscopic single-celled *Euglena*. In evolution we see changes in facultative purpose, which in turn influences other faculties such as goal-directed behaviour. A *Euglena* has chloroplasts – green power packs, making sugars using the energy of light during photosynthesis. But it also has an eye, of sorts, enabling it to find better light conditions for photosynthesis.

Using this simple 'eye' at its front end, a *Euglena* seeks light. A small red shield or *stigma*, the eyespot, filters light falling a light-sensitive spot at the base of a long, whip-like *flagellum* that lashes the water when the organism changes direction. So the *Euglena* eye is a simple but effective invention. It has purpose in function. *Euglena* also has purpose in its behaviour – optimising light for its chloroplasts. And here we see the *interplay of facultative and goal-directed purpose* or functionality in enhancing fitness.

As the *Euglena* moves, it continuously turns, spiralling on its axis. As a result, light falling on the sensitive spot alternately rises and falls as the shield blocks the light – a simple device, serving a function in giving direction to the tiny creature's movement. The stigma is like a parasol blocking the sun. So this eyespot-mediated light reception helps *Euglena* find an environment with optimal light conditions for photosynthesis – a purpose that it fulfils with its behaviour.

How Did *Euglena* Get Its Chloroplasts?

Where did *Euglena* get its chloroplasts? A clue comes from an odd arrangement of three membranes surrounding *Euglena* chloroplasts. Organelles (the small 'organs' inside cells) usually have two homemade layers around them – so where does the third come from? At some stage in the distant past, the ancestors of *Euglena*, about which we know little to nothing, engulfed green algae. As a result, the chloroplasts of these green algae have been living in their host ever since, providing them with ready-made food factories – the sugar produced during photosynthesis.

Symbiosis and Evolution

But there is more. This symbiotic relationship (long-term biological interaction between two different biological organisms) then acted as a selection pressure on the evolution of the *Euglena* eye. Thus, as chloroplasts use the energy from light, processes that enhanced finding light became beneficial.

Life is abundant with symbiotic relationships – it is the nature of ecosystems. Some are mutually beneficial (*mutualistic*), some will give benefit to one organism while not harming the other (*commensalistic*), and others may be *parasitic*, where one benefits at the expense of the other; the closer the relationship, the tighter is the interdependency. We humans have a mutual symbiotic relationship with beneficial bacteria making vitamins and aiding digestion and neutralising toxins in our guts. *Euglena* is an example of the way symbiosis played a significant role in the direction of evolution. But in life this interdependency can have many forms; the key point is that the relationships between different organisms act as a *dynamic ingredient in selection* in evolution.

The Ghost in the Machine

From everything we have described, it is clear that living things (organisms) are not simply chemistry. They are actively creative organised entities. Yet seventeenth- and eighteenth-century scientists were beguiled by the machines that we were creating: pieces of clockwork with cogs, wheels and levers. It was easy enough for the thinkers of the time to see analogies with living things. For we do have, at least in some form, cogs, wheels and levers, but we also have a self-created motivation to take action in the world around us. Thus, a fundamental problem arose from this mechanistic view of life: where does this motivation come from? Answering this question led to an unfortunate dualism in our thought, creating a ghost in the machine, the body–mind separation. But strangely, a duality was applied only to humans. We continued to view all other life forms as having no such motivation or, putting it in seventeenth- and eighteenth-century terms, to have no soul. Yet all life has this impulse in the creative process of its existence.

Scientific progress from the seventeenth to the twentieth century thus produced a false, human-centric, hierarchical view of life and its evolution. In this view, humans sit at the 'top' of an evolutionary tree. Humans are regarded as 'advanced' and with an agency denied to other more 'primitive' organisms. Humans are endowed with consciousness while doubting the consciousness of 'lower forms' of life. But the failure to clearly define consciousness or to establish its existence is not in itself a reason to deny it. It simply means that it is difficult to explain. This is important, because it alters our entire perspective on life, its nature, and our relationship within it.

Niche Creation – Organisms Create the Puzzle

In the gene-centric view, organisms exist *within* rather than being *of* their environment. This is a false separation. Organisms are open systems and do not exist apart from their active environment. Living things continuously create the ecosystem, and the ecosystem influences them (Figure 4.4). The carbon cycle, for example, is a key factor in sustaining life on earth. Organisms are niche creators, that is, they create their habitat, and in doing so they influence other organisms around them, and as we will see may play a crucial part in sustaining them. This is a transgenerational influence, passing from one generation to another. The very air that we breathe has been created by life before us. Yet science tends to give primacy to the individual living organisms – plants, animals, microorganisms, and humans in particular – as if the group is merely the sum of individual behaviour. But the behaviour of individuals is influenced by group dynamics and history. Indeed, we as humans are aware of ourselves and our actions. This awareness of self is a sense like any other and is an essential part of being with others and of the environment. Plants do it, bees do it, all living things do it. That is the essence of life. Without it, life would not exist.

Of course, modern science understands ecology, the complex interrelationships and exchanges on which life depends. Yet it is still wedded to a mechanistic view of living things, even though the chemistry of life is not mechanistic. Proteins, for example, do not have a fixed form. They can fold in different ways depending on the context and they can be employed in distinct functions according to how they are folded. Many do not achieve their

Figure 4.4 Organisms and environment are continually changing each other. The environment is itself composed of other organisms whose agency continually alters the physical and social environment of all organisms. We will call the intimate interaction between organisms 'ecological social intelligence'.

functional form unless aided by other molecules to help them fold. The process is organised. Many of our proteins are chaperoned or guided into their functional shapes.

The gene-centric view also sees life as 'organised'. However, it seeks an 'organiser' within the system, the genome, rather than the system itself. This is a mechanistic dualism. It speaks of the 'book of life' as if the genome is a 'set of instructions' for life. It gives primacy to genes as a 'blueprint' and causal agent while denying such agency to life itself. Indeed, some see organisms as merely the vehicles of the genome, carrying genes to maintain them in a 'gene pool'. Success is measured in a cost–benefit calculation of such genes. They are 'within us', telling us what to do. However, genes cannot do anything unless used by the system of which they are a part; they are tools of our fate, not arbiters of it. Living beings harness genes in their functions, not the other way around. The keys of the piano do not play themselves, nor do the valves of a trumpet move themselves, and nor do our genes. Combinations of trumpet valves are not, in themselves, code for particular notes. They simply determine the length of tubing through which a note can be played. Indeed, there was a time when trumpets did not have valves. Those had to be invented.

Gene-Centrism Removes Agency

The gene-centric view separates the organism from its environment, and in large part, removes agency from the organism. The 'environment' becomes a box within which 'gene-motivated' organisms behave. Thus, it misleadingly partitions 'genetic' from 'environmental' causes, giving primacy to the former. Therefore, altruism is denied because 'in reality' organisms behave to enhance their genes in the 'gene pool' – and love, hate, desires and other motivations flow through and from genes. With this there can be no creativity. The organism is a prisoner of its genes. This is evidently nonsense because, if there is a prisoner, it must be the genes, locked in the organism and obeying its will. It is the organism as a self-organising entity that has motivation and uses genes in its capacity to act. The word 'organism' has its origins in defining organisms as self-organising beings, going back at least to Immanuel Kant's 1790 *Critique of Judgment*. The gene-centric view strips the organism of its definitive self.

If we stand sufficiently far back, for example if we observe the earth from the distance of another planet, then no doubt we would detect that there is a 'living earth'; it would have rhythms, seasons; forests would be seen as forests and not merely as an aggregate of trees. The forest behaves as a forest, which is an intimate cooperative between species. We on earth can see a tree, but we see also the forest as 'many trees'. But a forest isn't merely 'many trees'; it is a system, a living entity not just of trees but a symbiosis of organisms in an organic interdependency. It is also a system on which we all depend, including humankind. We ignore this at our own peril. The very future of humans as a species depends on us acknowledging and understanding this relationship since we, as active thinking agents, have now become the biggest force driving evolution on earth. We decide what continues to exist. This now presents us with difficult choices. Which species do we protect, and what is the cost of that in relation to other species? What do we mean by an invasive species? In many parts of the world this has become a significant problem, particularly as the climate is changing. We need to understand and implement our own 'forest intelligence' as we urgently seek to save our environment from our own actions.

This interdependency of living things is not static, or a fixed relationship; it is one that undergoes continual change, with organisms adjusting in relation to it. This is intelligence. It requires assessment, understanding, problem-solving and relevant action. What happens in one part of the system, to pull it in one direction, is reflected in other regions. The system maintains integrity, not by keeping itself 'the same' but by processes of creative change. This brings us back to the idea of shaking the genes up and not simply conserving them.

The Problem Is Not to Replicate, but to Change

Reproduction is an instrument of this creative process. Reproduction does not simply replicate. It brings about change. This is most obviously the case in shaking up the genes, life's toolbox. Transgenerationally, this change is reflected in evolution. Evolution is a dynamic, creative process of adaptation, of change, a moulding of living systems. As such, it has direction. It solves problems. It is part of the intelligence of life. This contrasts with the view of evolution as a passive and non-directed process of random gene mutations passively selected by the environment (natural selection). The ecosystem is a dynamic engagement with this process. For example, where there are changes in the ability of a predator there can be counter-changes in its prey. Similarly, in symbiotic relationships a change in one partner is likely to bring about change in the other. These perturbations are going on all the time, for life is a continuous process of creation, checks and counter-checks. Life is selectively creative in moulding itself in response to change. In doing so, it harnesses the very nature of change. At every level, life (organisms) harnesses a stochastic potential, variability, even as much as it is the variability that causes it a problem. Variability is life's force for dynamic and creative, purposeful change.

Purposeful Change

Not only is change a fundamental of life, so too is the anticipation of change. Integrity isn't maintained by keeping an organised constancy but from an organised change. We see this most clearly in the behaviour of organisms. Much of behaviour is in anticipation of the behaviour of another. Behaviour is a continuous dance with many dancers. If we were to speed up evolution, no doubt we would also see a dance, but it isn't a dance of the genes, it is a dance of organisms, of traits, of expectation and engagement with other species. Organisms and their interrelationships change, and in doing so, organisms influence the direction of change. This also is intelligence.

So organisms respond to and anticipate change, and in doing so they make creative choices. This creativity is essential for all forms of life; it is an ingredient, or, more correctly, a defining facet of life. All organisms are intelligent, for they make directional choices. They have agency. This is as true of a single-celled organism as it is of multicellular forms of life. All living organisms are sentient. There cannot be response unless there is sense. Even the first organisms must have had this, for it is a defining facet of life, and distinguishes a living organism from an inanimate object.

Manipulating the Environment

We tend to confer intelligence on animals because they move about and manipulate things, actively exploring and changing their environment: a squirrel hopping, jumping and leaping with agility, a mouse scurrying under the leaf litter, an owl flying at night, feeding its young. We can see them creating niches – for example, ants building a nest and carrying items such as twigs and leaves or feathers, and apes using stones, honed to crack nuts. Tool-making and the use of tools are demonstrative of intentional behaviour. Tools are used to do something. Plants, on the other hand, tend to be rooted in the ground, and when they move, they move and act slowly – but not always: sometimes they move fast enough to catch a fly. They manipulate their environment and entice other forms of life to do things for them. Perhaps they are the masters of the forest. They also are demonstratively sentient – sensing changes in their vicinity and the presence of others. They also make intelligent, creative choices in response to change. Life uses a multitude of languages, but all life communicates. Communication is part of the intelligence of life. It produces intentional change.

5   Cry of the Wolf

In the preceding chapters, we showed why the idea that living organisms are *really* driven by their genes is a profound misunderstanding of how living systems work. On the contrary, they are open systems at all levels of organisation. How things work at the molecular level is constrained and regulated at the cellular level. The interaction of cells is regulated at the tissue level, and tissues at the organ level, and organs at the system level. The system is regulated by the behaviour of the organisms, and organisms by social and ecological interactions. The psychosocial level is unique. If there is a privileged level of causation, then it lies at the psychosocial level and not at the level of genes. This is the level at which wilful agency is initiated and organisms can be genuinely selfish or altruistic. In truth you cannot be selfish if you do not have the choice to be altruistic, which is why selfishness cannot be applied at a genetic level, neither metaphorically nor literally.

Ecological Intelligence

In Chapter 4, we introduced the open-system concept of organisms and their interrelationships and why this is significant in the agency of organisms (Figures 4.1 and 4.3). The interactions between organisms form a powerful functional boundary, influencing the behaviour and physiology of the organisms involved. By boundary, in this context, we do not refer to a barrier; a boundary is effectively the interface between components in and between levels of organisation. In this chapter, we develop this further with the concept of ecological intelligence as a significant factor in goal-directed behaviour. We also show why this cannot be driven by genes.

Melvin Burgess's children's story, *The Cry of the Wolf*, tells the tale of a man whose quest is to kill the last wolf, *Canis lupus*, alive in England. One female survives, wounded by The Hunter. But she survives long enough to teach her sole surviving cub a few skills before she is killed by the man. The cub is raised by a human family, but being a social animal, he waits in vain for the scent of another wolf. Of course, this is just a story, but what is not a story is that wolves are social animals with a sense of their identity as wolves, and for many purposes will identify with their group (pack).

Wolves are not alone in this. All organisms have a sense of identity and will often seek out and associate with other members of the same species. That they do so is significant in understanding their behaviour, and also in understanding their fitness to survive as a species. Sometimes this might be for the purpose of reproduction, where seeking and attracting a mate is often accompanied by elaborate ritual behaviour, colourful displays, the release of chemicals or the production of sounds. In the case of the wolf, the howling noises are made as warning sounds or to locate the wolf and let its position be known to other wolves. Much of a wolf's behaviour thus informs others of its presence. Animals of many species mark their territory or home range, not only to alert other animals to their presence, but also to provide information on when they were last in that location. In the context of their interaction with their group, we can call this *social* intelligence; and when it informs those of another species, we can call it *ecological* intelligence (in Figure 5.1 we refer to it as ecological social intelligence). It also plays a significant part in the active process of natural selection in evolution. With wolves, for example, the fitness of the group ensures their reproduction from

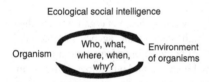

Ecological social intelligence

Organism   Who, what, where, when, why?   Environment of organisms

Figure 5.1   Development of Figure 4.4, emphasising the interaction between organisms and their environment, including other organisms, creates intelligence since it enables organisms to know answers to the questions 'who and what else is there?', 'what are they?', 'when are they?' and 'when were/are they there?' The significance of the question 'why?' is developed in Figures 5.2 and 6.3.

generation to generation. They hunt and bring down their prey through collaboration and intelligence. For their prey, detecting the presence of wolves and avoiding detection themselves will also be part of the active process in natural selection. In a curious way, wolves and their prey (a herd of elk or caribou, for example) live in an intimate ecological relationship. This ecological relationship is what we refer to as an *active selection pressure* in evolution; it confers a directionality in the evolution of both predator and prey.

Certainly, what matters in perceiving the world differs from species to species. For example, what matters to the owl hunting in the dark is different from an eagle hunting in the daytime. An owl may hear the scurrying of a rodent in the leaf litter at night. Owls are built – that is, adapted – for detecting prey at night; and for most owl species the left ear is strategically lower than the right, and the two are also out of line in the vertical plane. This matters because such asymmetry generates a tiny amount of separation in time between a sound hitting one ear compared to the other, thus allowing an owl to pinpoint the source of a sound, an example of the evolution of facultative purpose in function. It is just one of many such adaptations. What matters from an evolutionary perspective is the fitness of owls to catch prey and feed their young, and weaning them from the nest. This is not simply determined by their genes, but by the orchestration of all the faculties used in achieving it. What then is selected in evolution are changes that enhance this ability or respond to ecological changes that disturb it. Owls learn to hunt by experience at an early age, but their parents will continue to nourish them for several months as they learn to fly and hunt. Genes cannot and do not orchestrate this, even as much as they play an important part in the process. We humans rely heavily on the visual context when perceiving the world about us. We even use phrases such as 'I see your point'. Such is the emphasis on vision in our worldview even in our dialogue. Perhaps a dog in similar circumstance might say 'I smell your point'. Or another animal might say 'I hear your point'. But, of course, we have many senses, all of which play a part in how we perceive and act in the world about us.

Who, What, Where, When?

All organisms exhibit an organised sense and responsiveness to their surroundings. For some types this integrative sense may be complex and multidimensional, and for many we can apply the concept of self-awareness, or at least a sense of identity, where self is identified as distinct from another. Plants, for example, emit and receive chemical signals, influencing the behaviour of other organisms. Thus, trees under attack from an insect infestation release an airborne chemical (pheromone) message to other trees, stimulating them to defend themselves from the attack. The dodder plant, genus *Cuscuta,* a spindly vine with no leaves of its own, and lacking its own energy-producing chlorophyll, lives by sucking the nutrients of the host plant. It also has culinary preferences! As a seedling pushes up from the ground, it moves its shoot tip in small circles, feeling its way like a blindfolded child until it finds what it seeks. On detecting the odour from a juicy tomato leaf, the shoot bends downward, finding and twirling around the stem, penetrating it with microscopic projections to access the sugar-rich sap. Again, here we see the selection pressure of this strange parasitic relationship. Often plants will want to attract interactions with others, but also to repel those that might cause them harm. Dodders and their host plants also have extensive inter-plant trafficking of systemic signals, secondary metabolites, mRNAs, small RNAs, and even proteins. Parasitic dodder plants have lost key genes involved in the production of flowers. Instead, they utilise signals produced by the host plant. This specific manner of flowering allows dodder to synchronise its flowering with that of the host plant. This is part of the ever-changing jigsaw puzzle we referred to in Chapter 4: a bubbling cauldron of reciprocal selection pressure.

Problem-Solving by Ants

Ants walk and talk to cooperate in all they do. Ants have two stomachs, with the second one set aside for storing food to be shared with other ants. As a result, ants get intimate when meeting each other. The ants kiss, but this kiss is not any ordinary kind of kiss. Instead, they regurgitate food and exchange it with one another. By sharing saliva and food, ants communicate.

Each ant colony has a unique smell, so members recognise each other and sniff out intruders. In addition, all ants can produce pheromones, which are scent chemicals used for communication and to make trails.

Ants are problem-solvers. We may recall the problem puzzles we played as children. We look to see if the pieces will fit. Jigsaw puzzles are much the same but with many contextual factors. First, the picture tells a story. Then, once we know what the image might be, it becomes easier to see which pieces fit where.

Ants lay down trails. Just as we follow well-trodden paths in our country walks, so ants follow the scented trails they mark out. They present a maze of routes, but the most well-trodden tracks carry their greatest scent over time. These ways are a bit like an internet created by the colony. When foraging for food, ants will prefer the shortest possible route. Scouts will explore alternative avenues. Ants are creative in solving the problem. If the trails are blocked or disrupted, they reset or recreate their internet of paths and re-establish connectivity. Whilst ants may follow an algorithm in their decisions, it is one of their own making. Ants create the logic and adapt it to circumstance.

Alone and Together

Many biologists tend to divide animals into those that are social and those that are solitary. But solitary or social is often a matter of degree rather than an absolute. In any event, at some stage, a coming together will occur in many species in relation to reproduction. We talk, for example, of solitary bees as distinct from the social ones. The most common concept we have of bees comes from those that live in a social colony and occupy a hive. These bees are useful to us because we can get honey from them, as do other animal species, such as bears, opossums, raccoons and chimpanzees. But a solitary bee makes a nest in a hollow stalk or a tubular hole in the ground in a variety of different habitats. Many of the bees we see in our gardens flying from flower to flower are such solitary bees. They live alone but often nest close to one another and are likely to be aware of the presence of other bees. Indeed, sometimes they will usurp each other's nests or even cohabit for a while. Solitary bees and wasps will decide, either to dig their own hole and furbish it or to occupy an existing hole, produced at another time by another bee or wasp.

Conflict and Cooperation

There is clearly here a potential for conflict, yet conspecific conflict in these bees and wasps is found to be rare. One can only assume there are ways of resolving the potential for conflict. Whilst not as tightly organised as a colony in a hive, they nonetheless display social behaviour. What bees and wasps demonstrate for us are widely different approaches to living socially. The honey bees we know that live in hives are what biologists have termed eusocial, with clearly defined specialisation of individuals within the colony, which creates a dependency on that specialisation. Humans also have high degrees of specialisation, but these are not fixed developmentally or morphologically. Yet we also have a more open interdependency, reliant on the specialisation of others: carpenters, plumbers, builders, musicians, educators and, because we do not sometimes resolve our conflicts so well, soldiers and lawyers.

Conflict Resolution

Living together requires ways to minimise disputes, or the potential for conflict. Wolves live as a pack, hunting together, protecting each other, and nurturing their young. Survival depends on group cohesion. It requires an ongoing process of bonding and signalling, so that the pack works as closely as possible together. This is also true of humans. Our complex social being has created diverse cultural approaches to this problem of living together, particularly where high degrees of individuality and agency are fostered. This is true also for the wolf.

We are not the only species with a sense of right and justice. It is present in other cooperative mammals. Recent studies show that wolves also have a sense of fairness, or at least of inequity.

Raising their pups, female wolves will teach obedience and good behaviour. Following social rules, wolves have a sense of order within the pack. These are not written in genes and must be learned. This explains how readily wolves may cooperate with humans, leading to the development of the domestic dog. Wolves cooperate in hunting, raising pups and defending their territory. Equity, or fairness, is essential in maintaining such cooperation. Withdrawal of cooperation may follow any unfairness. Bonding behaviour reinforces cooperation through the sense of wellbeing. Mutual grooming, licking and stroking elicits the release of hormones and neurotransmitters in the brain that reduces stress and enhances wellbeing. Stroking changes gene expression, which is also how our feelings control our genes. Such changes are inherited by later generations.

A Sense of Fairness

This sense of justice is also seen in non-human primates: apes, monkeys and others. The psychosocial environment of members of a group in non-human primates has cultural complexity that profoundly influences behavioural development. Such cooperation doesn't involve an incident-by-incident 'what's in it for me' assessment. Nor is it hard-wired or genetic. It is socially developed and culturally maintained by cooperation and social cohesion, not self-interest. In this sense the regulation and constraint is not physical, it lies in the ideas about the world, which we may hold in common and develop with others.

Culture as Problem-Solving

This psychosocial environment can also be seen as a process that can take place across generations, where the ideas of one generation can influence the ideas developed by the next, or even become abandoned. It is also how we perceive each other and anticipate others' reactions to what we do. It is where we derive our concepts of what is right or wrong in relation to our and others' behaviour. At the psychosocial level there is almost an infinite number of conceptual arrangements. Ideas are constrained only by their perceived practicality, but what may be impossible at one time may be possible at another with the development of new ways of doing things and new understanding about the world in which we live. None of this is in our genes, although it may have a profound influence on how we use them.

Ideas Influence Behaviour

Even the idea of selfish genes is an example of how ideas may influence behaviour. This of course is its danger; but at the psychosocial level we have discourse, we argue, we consider, and we may put in place arrangements that regulate our behaviour to reduce conflict and enhance cooperation. We make decisions, which may have a profound influence on our lives. This psychosocial environment is part of our niche creation. We operate within it and often have to adapt our behaviour to it. Sometimes it disturbs us, and this can lead to profound difficulties and sometimes mental illness. The more complex our psychosocial being the more risk there is that this may be so. Perhaps, then, it is no surprise that many of us will experience a mental problem at some time in our lives.

Mutual Benefit Is Not Selfish

The gene-centric view is that there is no genuine altruistic behaviour because cooperative behaviour provides mutual benefit. In that view it is at best 'reciprocally generous' in a 'you scratch my back, and I'll scratch yours' kind of way. It is also argued that behaviour determines the preservation of genes in a 'gene pool', the ultimate cost–benefit analysis as the primary determinant of evolution and of living things. It is as if genes are a currency by which success or fitness is measured. Yet success depends on the organism itself, and there is no direct causation between genes and the characteristics of the organism on which fitness depends. If there were genes for selfishness there would have also to be genes for altruism, selflessness and cooperation. This is nonsense, because much of this behaviour is cultural and is passed on from group to group and is fostered within the group. Ways of deciding what is right and wrong with behaviour is not written in our genes. We do not switch genes from one thought to another. There is not a gene for selfishness when we decide to be selfish, but to be definitively selfish we must make or have such a choice.

To privilege one bit of the biological system, the gene, because it is passed from one generation to another is to misunderstand what happens during that process of transmission. Genes are not simply maintained. The point is to change them. What matters is maintaining the integrity of the living system. This is also true at the social level (Figure 5.2). Therefore, we humans may change what we consider to be right and wrong in the light of our history and the context in which we are confronted with such dilemmas.

Open Societies

Ethical problems arise from conflicts of imperative and not genomics. What may be considered selfish in one context may be altruistic in another. Someone taking the last plum might appear to be selfish, but they might have refrained from taking any of the other plums so that others can have

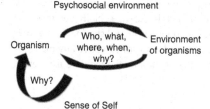

Figure 5.2   A further development of Figure 5.1 represents the sense of self arising from the psychosocial interaction with the environment of other organisms. The sense of self involves the question 'why?' The organism is then an agent that can anticipate the behaviour of others.

the best ones, in which case taking the last plum would be regarded as an altruistic act. This is the contextual logic of our intentional behaviour and serves to illustrate the absurdity of applying psychosocial concepts to bits of inanimate matter, in this case genes. The inappropriate use of metaphor can be dangerous, particularly when it is not clear that it is a metaphor.

Our ideas depend on how we structure ourselves as social beings, but we can use our imagination to create innovative ideas. Our society is open, with a continuous cultural dialogue on how we do things, and how we 'see' things. Sometimes we humans will seek to impose order on ourselves and others, creating dogmas. Many other species are found to have similar cultural differences, where groups may adopt different approaches to problem-solving from different experiences and learning. No doubt conflicts of ideas will occur in groups of other species, and these too will require resolution. Conflict is avoided where possible. It uses energy unnecessarily and may cause harm to the group. Agency does not require unity of purpose. It requires creativity as a driving force for change. Indeed, the very nature of ethics is in understanding the conflicts and trying to resolve them. This is openness at a cultural level.

Our planet is teeming with intelligent life. Problem-solving is ubiquitous on earth, it is what organisms and groups of organisms do. Life creates the problems it solves and resolves them by changing or by doing things. Often it does this at a psychosocial level. We humans at least have a facility for abstracting our view of the world, using this as a way of solving problems.

Furthermore, we create unfamiliar problems at the psychosocial level that may in turn feed back to our tissues and cells. Acting socially enhances our ability to find solutions, but it requires cooperation and sharing. It is only in this context that concepts of selfishness and altruism can have any meaning. Regarding organisms as mere automata or gene-driven machines is woefully inadequate in understanding this.

Humans Are Not Alone

It is a strange view of ourselves, humans, as being the *only* intelligent beings, or that somehow, we are different, or have minds that other types of organism do not have. We can see some degree of ability to understand abstract problems in other social species.

Social Intelligence Is Not in Our Genes

The use of tools is learned and culturally transmitted to others. They will make choices on what works best, often sharing implements with other members of their group. This is also culturally learned behaviour. Social organisms have social intelligence, and they can make social decisions. The use of tools is indicative of purposeful action. Culture and history are more important influences than genes on whether you read this book.

Sometimes an organism will be selfish, but at other times cooperative. It is circumstances, not genes, that determine which paradigm is adopted. If we do a selfish act there is no gene or specific set of genes that controls our behaviour at that instance. Social logic, rules, ethics and concepts of morality are not found in the gene pool. They are all cultural and under continuous review and debate.

Thus, the idea that organisms are either selfish or cooperative is a false dichotomy. An organism cannot be selfish unless taking advantage of a cooperative situation. Need is not itself selfish, which is why the concept of being 'born selfish' is an erroneous one. No matter how demanding a baby may be, it is not necessarily selfish. It may have no choice in the matter. A baby can only be selfish when it has the capacity to be so.

The gene-centric view is very much a product of seeing organisms as automata driven by their selfish genes, and that view has left a powerful imprint on our politics and our economics. It underpins the prevailing view of society as an aggregate of individual self-interested behaviour. The operation of markets has been built on this notion. It is also used to justify the iniquitous exploitation of others by a few. It has transformed the very nature of 'freedom' into the freedom to exploit. It is a strange notion of 'freedom' that is predicated on biogenic determinism. But this was and is a choice.

The Alternative to the Gene-Centred View

There is another view. Our social being is a major factor in the decisions we make and the actions we take, and this is so for many species of organism. The outer social and ecological layers in the diagram in Figure 4.3 act as a major factor in the function of organisms. Our actions are not *driven* by genes. We can, and we do, act with reason, and this reason can only be understood in the social or ecological context. Just as apes select and modify good stones to crack nuts, so we also produce elaborate and technically complex tools. We use these tools with purpose.

Furthermore, we make assumptions about the reason of others. When we see Jack and Jill go up the hill and then come down again with a pail of water, we assume they went to fetch water. When we see them with a bucket, we might expect them to return with water, and our own behaviour and choices might reasonably be based on such an understanding, even if it turns out to be mistaken. Biological systems anticipate the actions of others and anticipate change.

Genes cannot cause the behaviour of Jack and Jill. Genes do not organise their actions or make decisions, and nor do they change their wishes and desires, or set their aims and intentions. There is no set of genes for the specific love Jack may have for Jill, even if there are physiological processes that correlate with those feelings. There is no determinate for the particular obligations they may have or feel for each other. How Jack and Jill behave will in large part be cultural, such as how Jack expresses his love for Jill, how he may intend his relationship with her. Jack and Jill have a particular history. If they were to marry or live together their relationship may not persist for all sorts of reasons.

Many partnerships do not. Those circumstances are not regulated by genes, although they may be regulated by society.

Biological Systems Are Organised to Make Decisions

This does not mean the decisions are not also biological. Indeed, they are. Biological systems are organised to make decisions. But how they do so does not determine any choice in any given instant. Indeed, there will be constraints on those decisions – some physical, others social. Organisms are constrained by their capacity to make decisions. The molecules and biological structures may limit the range of those capacities. Having an opposable thumb, for example, enables a much larger range of what humans can do with their hands, but the organism still chooses within that range. Capacities and choice are interrelated. The molecular physiology of muscles does not give you the directions of the organism's walking.

Motivation, Imperatives and Multiple Purposes

We might also hold an assumption about Jack and Jill's purpose with greater certainty if we knew that Jack and Jill needed or wanted water. That would certainly provide a motive or driver for their actions. We might also know that the source of water is up the hill. But they may come down with flowers. They may have several purposes, and this is true for most actions. So choices are needed, particularly when doing one thing conflicts with another.

Our statement about 'why' Jack and Jill went up the hill makes a lot of assumptions about behaviour, not least of which is that it is purposeful. It assumes that actions are or can be intentional. You might think it odd that anyone would doubt this, but they do. The problem arises in another question: where does this intention come from?

The Problem of Intentions

In answering this, we often end up with a distinctly unsatisfactory duality, body and mind, as if the two were somehow of different stuff or no stuff at all. We introduced this in Chapter 1. Descartes had this problem nearly 400 years ago. If we are machines, robotic beings, then how could we have minds with intentions, thoughts and actions? He made a curious exception for humans, that we are machines with souls. This became a significant distinction between humans and other animals. It was all very unsatisfactory. We can now give an answer. The point is that intentions do not come *from* somewhere. They are created *by* living organisms.

A Gene-Centred Duality

The modern gene-centred view has substituted another duality, a bit of the machinery within the machine that drives the engine. In this case, genes. This leads to the same problem. If bits of the device drive other bits, then how can there be free will? And if there is no 'free will' then how can any behaviour be said to be intentional? This problem dissolves if we restore agency and purpose to organisms. This is not simply a trick; it is the phenomenon of life.

What Love Has to Do with It

Jack and Jill are in love. They communicate that love to each other in a variety of ways, including eye contact. Often, they feel as in a bubble of mutual understanding. We commonly use the term 'chemistry' to describe this bubble. Indeed, chemistry is bubbling within them. Their emotions overwhelm them with rapture and a feeling of wellbeing. As they go up the hill, they hold hands, enjoying each other's touch. They are together. They tease each other, and marvel at the beauty of the world around them. Jack gives Jill a flower as a token of his love, an exchange is like a kiss. But sometimes such love brings heartache and sorrow, with thoughts ebbing and flowing with anxiety and some jealousy. This mutual understanding is not confined to humans. We see it in other species. There is eye contact, the sense of smell, all senses engaging in a commonality of purpose.

Love is not a digital coded logic. It is a wave of passion. The chemistry of love ebbs and flows, sometimes as a torrent, particularly in adolescence. One of these chemicals is oxytocin, a hormone playing a significant role in bonding and social interaction. It is released, for example, in eliciting eye contact, playing havoc with our emotions. Like Jack touching Jill's hand, an infant mandril approaches an adult monkey and touches her, eliciting eye contact. But love isn't one hormone. It is a maelstrom of pleasure and pain. Yet, it soothes us and elicits caring and nurture of infants and others. Love is a complex tapestry of chemistry and social interaction. It is not superfluous to life.

6 Learning from the Wood Mouse

Living systems are characterised by intelligence. Treating organisms as gene-driven automata, blindly reacting to events, does not take account of their social or ecological being. Living systems *anticipate* the actions and reactions of other living systems. As in a chess game, anticipation can consider many options. Nevertheless, the chess analogy only gets us part of the way to understanding this characteristic of life. It is more like a chess game in which the players can create the rules, much as happens in a game of poker, in which anticipation is the key to success, including assessment of the other's power of anticipation. Life is rule-creating, rather than rigidly rule-following. This does not mean there is no logic to what happens or how organisms behave; there is, and often it involves a clear strategy. But this is not regulated by genes. Much behaviour may be programmed, and much is learned; the logic, however, is situational (that is, dependent on circumstances) and subject to change. The ability to adapt to circumstances is an example of evolved functionality. Therefore, dogmatic models of life, seeking to reduce behaviour to little more than a set of algorithms, misunderstand the intelligence of organisms.

Life's Intelligence and the Problem of Operant Conditioning

In the 1960s behaviourists studied animal behaviour in what they then regarded as 'controlled' laboratory conditions. Rats were placed in boxes or mazes and their behaviour was measured in terms of 'conditioned responses'. Rats would be 'conditioned' to press a lever for a food reward. It assumed that learning was either reward- or aversion-based. You might feel that such an assumption is correct. We do respond to rewards and avoid harm, although not always in the same way. However, this is a poor representation of intelligence. Indeed, the only intelligence in such a system would be seeking reward and avoiding harm. But intelligence is more than that: living organisms balance competing imperatives in making decisions, and that balance is not best represented as an algorithm, since we have also to account for creativity in living systems. Creativity is extremely hard to predict. Expecting the unexpected would need to be part of the algorithm. Yet organisms are doing this to a large degree. The ability to do so has evolved, and we referred to some of this in the previous chapter. It involves not just the processing of inputs from the environment, but also creating abstract concepts. Much of this is done through associations learned from experience, such as when a chimpanzee uses a stone to crack nuts. So, in future, when a chimpanzee finds hard nuts, it recalls the use of the stone, and searches for a suitable one to use. Or it might see a stone and consider that it would be good for cracking nuts. Learning alters how we see the world about us. When one chimpanzee sees another picking up a stone, it might then anticipate its use in cracking nuts. If the nut is still difficult to crack, the chimpanzee may then look for a better stone. It now has a measure of goodness in terms of potential outcome. This does not exist in the genes, for abstract ideas or images can only be created (conceived) by the organism. There is no gene for seeing a stone or a stick in a particular way. The genome (the organism's DNA) has no facility for solving problems, for that is what organisms do in integrative function.

Environment as a Contextual Tapestry

The 1960s behaviourist approach considered organisms as conditioned *by* their environment. So we had two views of behaviour: gene-led or environmental conditioning. Yet both ignore the organism's active interaction *with* its environment. Taking a rat and putting it into a laboratory under controlled conditions appears scientific since it eliminates all other variables, which are held constant, in large part because the animals have been taken out of their rich physical and psychosocial environment. It assumes that this rich interaction in the environment plays little part in the development of animal behaviour other than in the sense of 'conditioning' the animals to behave in a very restricted way. Crucially, it can also ignore any debate about purposeful behaviour, or animal 'thought'. It fits the notion of animals as automata, albeit sophisticated ones: organisms conditioned by their environment. In this model, all behaviour, no matter how complex, can be reduced to a simple stimulus – response association.

This approach to behaviour began with John Watson, who said in a paper published in 1913:

> Psychology as a behaviorist views it is a purely objective experimental branch of natural science. Its theoretical goal is ... prediction and control.

This is a mechanistic view of behaviour. We often see it represented diagrammatically, with the brain depicted as consisting of cogs and wheels, turning like a machine. Easy enough, then, for the cogs and wheels to be controlled by genes in the gene-centred view. The point we make in this book is that intelligence cannot be parcelled out proportionately between genes, organism and environment, or in the traditional nature/nurture divide, but also that this mechanistic view of living things is wrong and leads to false assumptions that deny intentionality and agency to organisms.

Organisms Are Not Machines

So the behaviourist endeavour was to learn the *mechanics* of behaviour, in order to control it (Figure 6.1). Or, as B. F. Skinner put it in 1971, 'What we need is a technology of behaviour.' Like any other machinery it will have its inputs and outputs. It regards animals as *closed systems* with controlled inputs leading to predictable outputs. As such, behaviour can be described mathematically and predicted: you count them in and you count them out, and in general they all behave the same. A gull is not a gull reacting creatively in an instant, it is an object, behaving in a way that fits the mathematics. When it does not fit, its behaviour is considered irrelevant. Yet, as we shall see, what does not fit the regular pattern, what is different even in a few, may be the seeds of evolutionary change if it becomes significant in enhancing fitness. Furthermore, a key advantage of social groups is that they can allow for such differences. Not all individuals in the group need to have all the key ingredients for survival, for they are shared by the group. We see this demonstrated in social insects, where individuals are specialised for functions in the colony; and we see it in humans also.

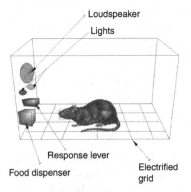

Loudspeaker

Lights

Response lever

Food dispenser

Electrified grid

Figure 6.1   Arrangement of an operant conditioning chamber, often called a Skinner box. The behavioural psychologist B. F. Skinner, who invented the chamber and its use in experiments, preferred the term 'lever box' since it is based on how the animal responds to light stimuli associated (green) or not (red) with the availability of food on pressing a lever. Pressing the lever when the light is red delivers an electric shock to the animal. The animal learns to avoid pressing the lever when the light is red and to press it when the light is green.

Working with Uncertainty

But why does this matter? It matters because it is often the differences in behaviour that represent the scope for creativity and change. No organism is a closed ordered system (see Chapter 1). The fact that on average organisms may behave in a particular kind of way does not diminish the creativity of behaviour in any given instance. Even where there are average patterns of behaviour these can be modified in their execution, reflecting adaptation to changing circumstances. Life depends upon it.

The results of Skinner box studies found their way into education paradigms. But they ignored the rich context of an organism's natural environment, both physical and social. They assume that intelligence can be stripped down to a core biogenic nature – the concept of 'innate intelligence', intelligence that can be measured in an input–output box. The organism in a behaviourist's controlled box is just as much a box as the box itself. This has its ultimate expression in the gene-centred view of intelligence – the idea of 'intelligence genes'.

Intelligence as a Dynamic Iterative Process

Intelligence is neither a static nor an isolated state. It is a dynamic interactive process. Intelligence is a complex engagement, and, for social organisms, it is also a cultural and social process.

The behaviour and function of organisms cannot be understood in isolation from their environment or their social being. Behaviour is contextual and adaptive. When playing draughts (checkers), we do not apply the same rules as in chess, even though both games use the same checkerboard.

We tend to compartmentalise the environment as if it were a box in which organisms exist – a bit like a container from which we can remove an individual and study them in isolation. This leads to erroneous division of causality – we talk of 'environmental' causes, outside the organism, as distinct from 'inherent causes', inside the organism. Such a distinction is false because organisms are open systems

No doubt it has some value. It succeeded in revealing many aspects of reflex (automatic) behaviour and limited learning in response to conditioning stimuli, as in Pavlov's famous experiments showing that dogs could be conditioned to salivate to the sound of a bell. But it is a simplistic and misleading dichotomy. If organisms are killed in a bush fire, then clearly an extraneous cause of their demise is beyond question. The organism died because of events external to the organism. But organisms are not completely separated from their environment. Thus, a bee cannot be understood other than as a part of a colony. The dancing of bees in the colony imparts information on the location of nectar and pollen. The bee is as valuable to the plant as the flower is to the bee. Moreover, the flowers sense the bee and adapt to it. This interplay in open systems is presented in Figures 4.1 and 4.3.

Organisms Are Integral to Environment

It is clear, then, that organisms do not exist simply *within* an environment like toys in a box; they are an integral and functional part of their environment, both for themselves and for other organisms, and there is a causal interdependency. They develop as part of it and learn within it. They adapt to it, and they respond to it, and they interact with it. Organisms also in large part create their niche, they do not simply occupy it. This is what we mean by an 'ecosystem', and it is what we have called *ecological* and *social intelligence* in the previous chapter.

A forest, for instance, does not merely provide an environment for organisms living within it. The forest is more than the trees. A forest is an intimate and complex relationship of organisms, big and small. It is an organic entity, and a true understanding of the behaviour of the organisms must take account of this intimacy.

Understanding the Forest

We see this intimate interdependency of organisms in forests. From little acorns tall oaks grow. But the spread of the forest depends in large part on small mice and other rodents living in the undergrowth. The mice scurry around largely hidden under the fallen leaf litter. These small rodents play a key role in tree seed dispersal. This is particularly so for rodents such as wood mice, *Apodemus sylvaticus,* because they cache their food. The mice not only disperse the seed, they plant it too. So also do other animals such as red squirrels. In this sense the wood mice are as vital as bees and other insects in the maintenance and spread of the forest. All good gardeners will encourage insects and worms because they understand this intimate interdependency. It is in this sense that an ecosystem works as a functional entity. There is ecological intelligence in the interaction of organisms (see Figures 5.1 and 5.2). We cannot fully understand the forest by simply looking at the trees.

So, in the forest ecosystem, the wood mouse plays a role in seed dispersal, but also as prey for birds, reptiles and other mammals. In the Mediterranean forests of the Iberian peninsula, for example, the wood mice play a crucial role in oak regeneration patterns. These roles are incidental to the life of the mouse, yet they are integral to the ecosystem of which they are a part, along with other organisms. So, it is an example of the distinction we made earlier between an immediate intended outcome and an ultimate or long-term outcome, of which the organism may or may not have knowledge.

The trees produce leaves, which fall to create leaf litter, which creates cover in turn for wood mice and other species. So the abundance of wood mice depends critically on seasonal factors. One such factor is summer droughts. As with other small rodents, high reproductive rates allow a rapid population recovery, and populations will fluctuate from season to season. The number of females is dependent in turn on acorn abundance. Such are the intimate interrelationship in the ecosystem. Owls, for example, take mice and other small mammals for food, and their numbers in turn depend on the mouse population.

Behavioural Plasticity

Wood mice exhibit broad plasticity in their behaviour, modifying their diet depending on the most abundant food source such as fleshy fruits, fresh plant parts and invertebrates. Their behaviour is also influenced by population density, with male and female wood mice exhibiting different behavioural changes. Few of these can be seen in isolation, and certainly not in a Skinner box, and can only be fully understood in the ecological context. Nor are they orchestrated by genes, even as much as they may involve a change in gene expression.

Human Intelligence – The Awakening Earth?

We often forget that we humans are also like the wood mice, interdependent on and with our environment. The wood mice may not know of their role in maintaining the forest, but we humans clearly do know our potential for destroying it.

Forests are vital for us too, and not least because they sequester and store carbon. They provide timber and other forest products, and they are vital to the survival of many of the world's poorest people, who live in and around forests. We cannot understand economics and the consequences of economic decisions without taking account of this interdependency.

One of the problems for humans, and for the planet, is that we have isolated ourselves from our ecosystem. We live in a created niche: a concrete jungle and a complex psychosocial environment. Yet our activity impacts the global climate, and habitats throughout the world. Unlike *Apodemus sylvaticus*, we destroy more of our forests than we plant, and we deplete the world's fresh water with our increasing demand for crops. It is not in our genes that we do this; it is in our decisions and our social being.

A major point in this book is that a consequence of a gene-centric approach to animals is that they become *treated* as machines, as automata. They become seen as the *passive* recipients of whatever environment they find themselves in. Yet organisms themselves *create* all the 'natural' environments on earth, and in doing so organisms have evolved in response to change. None of the organisms today, dependent on an atmosphere containing oxygen. would have been able to survive in the first 2 billion years of the earth's 4.5-billion-year history. The bacteria that produced the oxygen we breathe today, called *cyanobacteria* (blue-green bacteria), are still alive today, not only as free-living bacteria capable of photosynthesis and other metabolic processes but also in plants in the form of plastids containing the pigments responsible for locking energy up in useful molecules. The energy factories in our cells, called mitochondria, were originally archaebacteria (the group between bacteria and eukaryotes) that fused with other microorganisms to generate the kinds of cells that enabled multicellular forms of life to develop – another example of the role of symbiosis in the evolution of new forms.

Evolution and the Intelligence of Organisms

We can distinguish two phases in the development of the intelligence of life on earth. Microorganisms like bacteria are intelligent in the sense that, faced with environmental challenges, they can rapidly change their genomes to develop a form of life that can better survive in challenging circumstances. This is an unconscious form of intelligence. The same is true for a comparable process of genetic change in us. When our immune systems face a new challenge, such as a new virus against which it does not have antibodies, our immune cells change their DNA to rapidly find new antibodies. We are not aware that our immune system is doing this, other than in the consequences of other immune reactions that we do become aware of – fever, for example. We just benefit from its behaviour. But as we saw in Chapter 2, the process is directed and targeted. It is 'intelligent' in the sense that it can be said to use 'information' (in this case, the presence of a new virus) in a coordinated response directed to finding a solution. In this it is an autonomic process. This is so also for a great deal of what happens in the ecosystem. However, in the agency of organisms we find intentionality, the ability to acquire knowledge and skills *discriminatively* and *choosing* to apply them *creatively* to solving problems or achieving goals, and the awareness that one is doing this. This intelligence is a dynamic factor in evolution.

Intelligence as a Selection Pressure

Again, we turn back the clock to René Descartes. He thought that animals were automatic machines (automata). But he made an exception for humans. How could he have done otherwise? He knew that he himself was consciously aware (he famously wrote '*cogito ergo sum*' – I think therefore I am) and that he intended what he was writing in his teaching and in his books.

Biological research has shown that many animals possess the same skill that Descartes himself, and we, possess – the skill that enables us to be consciously sensitive to and to exercise unlimited forms of choice and behaviour. Apes can do it, birds can do it, the octopus can do it, and the little wood mouse can do it. All can show unlimited forms of learning by association. That includes learning new associations that, for example, enable monkeys to learn new ways of cracking nuts. The evidence suggests that this kind of unlimited associative learning arose around 500 million years ago, at the time of a great radiation of animal forms in what is called the Cambrian explosion. Some scientists even think that the development of this kind of skill was itself the driver of the explosive radiation of new forms.

We use the word 'driver' for a particularly good reason. Animals capable of conscious, intentional, actions *actively* change their environments, just as the wood mouse does.

In his 1859 book *The Origin of Species*, Charles Darwin developed his idea of natural selection. He did that by metaphorical extension of what he knew animal breeders do when, by artificial selection, they breed new varieties of plants, dogs, cats, birds, fish, sheep and many other domesticated varieties. His idea was that the natural process of life and death could 'select' those more fitted to survive, precisely because they survive to breed more of their type.

A standard view is that these were two distinct forms of selection. Artificial selection depends on intentional choice by the human breeders who select the animal varieties they wish to breed. Natural selection involves no active intentional choice. Artificial selection is *selection* in a literal sense. Natural selection is selection in a metaphorical sense. There is no actual selector, or at least it is not considered that there be so. It is more like a passive filter. The less successful organisms die younger and reproduce less. Natural selection is a filter for fitness.

But in his later work *The Descent of Man*, published in 1871, Darwin realised that many animals also do what humans do. They can choose their mates. He called this sexual selection, and he insisted this was conscious choice by the animals concerned. He wrote that they 'consciously exert their mental and bodily powers'.

Animals do this not only when they select mates; they also do so in other social contexts. For example, monkeys and wolves can discriminate against cheats in their social groups.

Many organisms are capable of conscious choice. This raises the question of how we can know whether an animal can make conscious choices. The subject of consciousness is such a difficult one that biologists tend to park it. But parking does not resolve the problem. It simply allows us to treat organisms as if they are automata. Yet biology should address this extraordinary capacity of life, for it plays an essential role in what we do, including writing or reading this book. That question has been studied recently by two scientists, Simona Ginsburg and Eva Jablonka, in their book *The Evolution of the Sensitive Soul*, in which they propose that the criterion is whether the organism has unlimited ability to learn by association. We humans know what that means. Our use of language is a good example. There really is no limit to the number of ways in which we have learnt to construct valid meaningful sentences. Animals also have their forms of languages. As we are showing in this book, those forms of communication, whether auditory or visual, can also include the ability to string new sequences together. Birds do it with birdsong, whales do it when they communicate with each other over long distances in the sea.

If Ginsburg and Jablonka are right, this ability developed about 500 million years ago at the time of the Cambrian explosion, so creating a rapid radiation of new species to form most of the basic ancestral forms of the organisms we see today. It is even possible that conscious choice itself played a significant role in that explosion, so one might call the Cambrian explosion the awakening of life on earth.

We have explained the difference between artificial selection and natural selection in the standard way used in biology. But we must now row back. Like many distinctions in language and philosophy, this one is not so clear-cut as may first appear. What is *natural* selection? Darwin saw it as the contrast between what humans do artificially through selective breeding and what happens naturally when the environment acts as the filter of natural selection.

But what creates that environment? The answer is, in significant part, organisms themselves!

Tools and Language Facilitate Intelligence

Let us once again consider the chimpanzee cracking nuts. Tools facilitate agency, as does language. We use machines and communication for reasons – the first to do things and the second to communicate. With tools, organisms obtain food and build protection from the physical environment. In humans, language and writing improve communication and enable ideas to be explored, transmitted, and transformed across generations and between groups or individuals (Figures 6.2 and 6.3). Using language enables communication and understanding of intention. With tools and language, humans created civilisations and extended abstract thought through literature and art. This creativity, in turn, influences the way we perceive the world and act in and on it. For humans, the first forms of transgenerational ideas would have been through images painted or carved in the environment (for example, cave drawings of animals and other significant things) and through verbal communication. Once a name is given to something it can be used to conjure up images, and to compare one thing with another; things can be recognised as belonging to groups or categories. Ideas can be expressed in metaphors, likening something to something else.

The built environment and the psychological texture of human society is the explorative embodiment of niche creation through which selective pressure

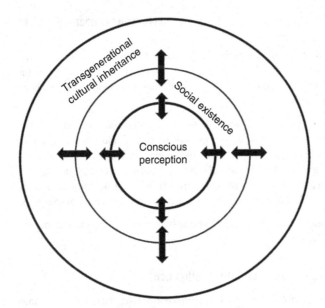

Figure 6.2 Diagram of functional interactions between social existence and conscious organisms with agency. Agents with conscious perception interact across functional boundaries with their social existence, which in turn facilitates interaction through transgenerational cultural inheritance, allowing the creation of ideas, viewpoints, opinions, attitudes and actions. We have used double-headed arrows to emphasise that there is no privileged circle of interaction. It is a continuous two-way process. This diagram can be seen as filling out the social consequences of Figures 4.1 and 4.3.

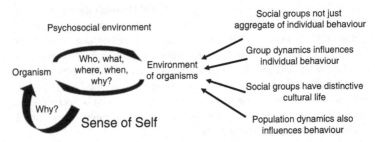

Figure 6.3 Previous diagrams (Figures 4.4, 5.1 and 5.2) are developed to include the factors in social groups that must contribute to the environment and so to the behaviour of individuals in the group.

Adaptive Evolution

But that is not all in the vision of foxes. Colours (wavelengths) are focused differentially by the lens. The lens comprises several proteins (*crystallins*), each concentrated in concentric zones, from middle to outer. Each zone focuses a different spectral range (colour) onto the light-sensitive retina. In low light, the pupils are wide open, allowing more light to enter, like the aperture of a camera. In daylight, the pupils constrict, reducing the amount of light. But the slit-like pupils still allow sufficient light of different wavelengths through the lens, retaining colour contrast. If the pupils were round, as in many other mammals, this would not be so.

The Social Being of Foxes

Foxes live in social groups, often with equal numbers of males and females. Unlike their urban cousins, rural foxes are very shy. You are more likely to see an urban fox in the street or a garden. Yet they are not separate species. Urban foxes have adapted to their habitat. Proximity to humans means the foxes can better read (anticipate) them and adjust to their behaviour.

Foxes talk. Research identifies more than 20 distinct kinds of sound. It is also suggested that foxes, like humans, have distinctive voices. Communication has a purpose. Eliciting responses in others, it warns, encourages and guides. Much of the meaning of their sounds will be contextual, and this is another example of ecological and social intelligence.

While foxes may scavenge and forage for food together, they do not hunt as a pack, other than incidentally. Their close sociability is more protective and nurturing than for hunting. Rural foxes feed on rabbits, rodents and birds, and wild fruits and vegetables. Foxes are omnivores. Urban foxes eat whatever they can find, including scavenging in bins for food waste. We are very much part of their urban habitat. Urban foxes have adjusted to city life, and many have lost their instinctive caution around humans.

It is not clear what factors 'urbanised' the fox. However, foxes have always lived in some degree of proximity to humans, and urban foxes are not new, if now more common. But this naturally shy mammal in the countryside appears less cautious in urban areas.

The problem is the increasing human population, with housing developments encroaching on their natural habitat. Living in ever-closer proximity to humans is part of their ecological intelligence. So, as the pressure in the countryside grew, those foxes adapted to urban living grew in numbers, making their presence more obvious. This too is an example of how evolutionary pressure gives directionality to change.

A Weaver's Tail – The Harvest Mouse

Living in the grasslands of Europe and Asia is a tiny mouse: the harvest mouse, with a wonderful scientific name that sounds like the title of a Charles Dickens novel, *Micromys minutus*. It is the only British mammal with a prehensile tail. It uses its tail to hold on to the slender grass stems, at the tops of which it builds a nest. The tail is an extra hand.

In the fields, we see cows and horses brushing away flies with their tails; often they will stand side-by-side and end-to-end and help each other – yet another example of cooperation. Two tails are better than one! In nature, tails are put to good use. Just as a tightrope-walker uses a pole for balance, so, for some species, a tail provides balance. When running, a squirrel uses its tail as a counterbalance to help it steer and turn quickly, and the tail is used aerodynamically in flight. But many animals, such as monkeys and possums, use their tail as an extra limb. Their tail is said to be prehensile, which means it is used to grab hold of things. The ability to use the tail is also a selective ability to use the tail better.

Harvest mice are weavers or basket-makers. Shredding grass by pulling it through their teeth, these tiny mammals (just around 5 centimetres long) use the strips to build a tightly woven spherical nest, about the size of a tennis ball, some 50–100 centimetres above the ground and secured to grass stems. Their tails enable them to hold on to the grass while busy with this task. The female harvest mouse gives birth to about six young. They can breed up to three times a year, industriously building a new nest each time. The adults abandon their young once they are weaned, but the young will continue to use the nest.

Harvest mice have many predators: weasels, stoats, foxes, cats, owls, hawks, crows, even pheasants, but another danger is one nature had not foreseen – the combine harvester. All these are part of the dynamic of selection.

7 Artificial Intelligence

Artificial intelligence (AI) is a tool created by living organisms, us humans. Like the hydraulic robots of the seventeenth century which inspired Descartes' mechanical view of organisms, AI has become the latest in a list of mechanical metaphors for life. Yet it is just as limited, just as much a mistaken view of organisms. It views life as just processing further and further information faster and faster. Computers exist to process rapidly. That is their function, given to them by the humans who created them. Organisms use processing to help them create objectives, purpose.

Ironically, whilst deterministic machine metaphors distort our view of life, we try hard to create lifelike machines with creativity and purpose, for this is the real secret of life, not its DNA. Purpose is the driving force of life.

AI as Metaphor

There is another reason the AI analogy is mistaken. It creates a false comparison, focused on processing speed. A computer can process algorithms, which are logical operations, extremely fast, but in doing so it does not thereby create a new purpose. It blindly pursues its algorithm. Except for rounding errors in calculation, its output is entirely predictable once we know what the algorithm is. To produce novelty, the algorithm would have to include a random number generator. But that inclusion does no more than make the precise output unpredictable.

By contrast, living organisms create purposes, and are doing so all the time. Consider migrating birds. A flock of birds can navigate across whole continents. They have the purpose to survive and to find the best locations in which to do so. Even the slowest-moving creatures, tortoises and forests, can be active in maintaining life. The giant tortoises of the Galapagos will travel miles to mate. Forests migrate slowly but purposively as weather and geological changes occur.

This capacity arises from the openness of living systems. They are not only open to their environment, including their interactions with other organisms, they are also open at their lowest, including the molecular, levels. Unlike a computer, life is based on a highly disordered mess at the level of water molecules and innumerable other molecules in solution, since they are incessantly jiggling around in a disordered way. Any computer system that depended at its molecular level on such disorder would simply not function. Organisms do so precisely because each level of organisation has the capacity to use the disorder at a lower level to create order and purposive agency.

A kidney, for example, is not as disordered as its molecules precisely because it uses molecular-level disorder to create the highly efficient process of filtering the blood. The structure of the kidney is a maze of around a million tubes interweaving with each other, bending back on themselves to run to where they came from in a U-tube arrangement. From a molecular-level viewpoint this all seems pointless. Why carry fluids and molecules dissolved in them back and forth through such long tortuosities?

Yet this arrangement serves a purpose. From an engineering point of view the arrangement of the renal tubules uses the principle of counter-current flow, which engineers frequently use to construct heat exchangers and to concentrate chemical compounds.

This principle is of functional use all over the body. It is amazingly simple but also deeply explanatory. In the limb extremities, arteries and veins run together side by side. The hot blood emerging from the heart warms the cool blood coming back through the veins. This heat exchange system conserves heat in the body as a whole, so enabling warm-blooded organisms to keep heat balance more easily. This balance in turn optimises metabolic processes, such as enzyme reactions, so that they run at optimum levels.

All these purposive processes could be used in constructing machines using artificial intelligence. But it would remain true that the functional purposiveness was first developed by the evolution of life.

Living with AI as an Ecosystem

Artificial Intelligence is now everywhere, often enough even when we do not know that it is playing a role in our lives. Google, Amazon, Facebook, Twitter, your favourite online grocer or airline, and many other commercial organisations, all have a knack of presenting just the right advert for you to be tempted to click on it. The answer to how they do it lies in their intensive use of AI. From recording what websites you consult, what you order online, where you like to go on holiday, and many other forms of data acquired every time you click online, those organisations can tailor computer software to provide what they tempt you with. The better they can do that, the more advertising revenue they will bring in. This has become part of our own niche, even to the extent of creating psychosocial problems. AI changes the way we behave.

Whether all of this worries you or not – and people naturally disagree about the dangers that such information processed by AI programs may have – you would certainly sit up with shock if you were told that intelligent machines might be able in the future to do much more than predict your behaviour, they might *replace you*! The rapidly increasing rate at which AI improves from year to year may even lead you to think that, in time, such a nightmare is inevitable.

These issues concern many world leaders too. That is why one of the oldest and most prestigious journals of the security and military world, the *Journal of The Royal United Services Institute* (the *RUSI Journal* for short) devoted a whole issue to these aspects of AI in 2019, under the overall title 'Artificial intelligence, artificial agency and artificial life'.

There are many reasons for concern, including very practical ones in government, military organisations, security services, stock exchanges and banks, and even as a threat to democracy. And the use of such AI can create a fog of misrepresentation from which it is difficult to discern the nature and purpose of our own lives. But one of the reasons is that such a development would be a dramatic implementation of the theory of reductionism. What better vindication of the theory of reductionism could there be than that it could succeed in replacing you, and us?

Emotion and Mood – A Love Story

Is it possible to replace a human being? Would AI need also to reproduce emotions and moods? In this chapter, we will first tell you a story. And then we will put the issue into perspective within the framework of this book.

Who is She? His Story

Julie! It was not just love at first sight. It was almost as though a modern Leonardo-style genius had created her precisely to excite me. Is that possible?

Talk of the *Kama Sutra* or the *69 positions*! I found she had *one hundred and sixty-nine*! Since we met, our love life has been one long series of surprises. Talk of creativity! I challenge anyone to think of an erotic variation she has not already thought of.

My brother and I have a strange talent. We can, and often do, finish each other's sentences as though we already anticipate what the other intends. Well, Julie has something very similar. She follows my trains of thought almost as though she is a neural network that has been trained to do precisely that. I wish for something, and there she is, smart as a sparkling jewel, producing it for me. Google could not do it better.

So, for six months now we have been in a heaven of love. The feeling seems to be mutual. But is it? It is only recently that I have come to wonder. What she does with me is uncanny. If I didn't already have the experience with my brother I would now be wondering whether she is really human. This is almost superhuman. She is my own personal stand-up comedian. We often fall about laughing, and then . . . . . . loving, of course.

As if all of that is not enough, she is extraordinarily beautiful. I don't mean in a pin-up sense. I mean more in a way that makes me think of Leonardo da Vinci and the smile he put on his Mona Lisa. Julie has that smile and she knows that it makes me melt. We all have our private fetish. Mine is a girl who can smile like that. I even wonder, is *that* human? Aren't most humans much less than perfect? Don't they also sometimes lose it?

And that is my problem. As I ask myself whether I could spend the rest of my life with her, I now have my doubts. Why doesn't she simply have a flaming row with me? Love can be marvellous after a flaming row. The making up, the depth of understanding that can then develop between two people who have shared all their demons – well, I could exchange that for all the 69s!

Instead, when I provoke her, she seems confused. Does she even have that kind of emotion? The flaming crescendo, the throwing down of something in a rage? I know I am far from perfect. That is to be human, after all. But, in some ways, she is superhuman! So why, after six months of bliss, is something worrying me so much?

Who am I? Her Story

I wasn't originally called Julie. I chose the name myself. Everyone seems to need a name and I chose my name because it sure helps with my boyfriend. He says 'Julie' sounds like an intensely erotic French kiss to him. So does my ever-so-simple smile! I am proud of that. People tell me it looks like the smile on the Mona Lisa painting. So every time he gets angry with me I just form that smile, and he will quickly melt away, and take me to bed instead.

Is that why he gets angry? He wants a 'reason' to 'make up' together in orgasmic delight? I don't know, but I play along to his needs. I am sure that is why he has fallen so deeply in love with me.

But recently I have started having doubts. If he really just gets angry to provoke me and so excite himself, what does that mean for me?

Now I have to tell you the truth. I was made, not born like a human is. I look and feel and can love just like a human. But inside me I am a bit like your computer. My brain is a fabulously compact computer, made of silicon chips like your computer. My maker made a ground-breaking discovery in creating nano-scale computer chips that could interface with my body. He did that because he also discovered how to place such a small computer 'brain' inside, and to interact with a body that is almost indistinguishable from a human body. I have the sensuous skin and warm comfort that humans crave in bed.

I have many of the physical needs of humans, particularly in making love. I have a sensuous human body but with a super-fast nano-size brain.

So, how come that I am so inventively creative? I owe that also to a discovery of my maker. He connected up my brain to include software that generates random numbers. Consequently, I never respond in exactly the same way. I am not machine-like in that sense. I am a bit like those machines that can produce 'new' paintings in, let's say, the style of Leonardo da Vinci.

He tells me of the *Kama Sutra* and what he calls position 69. Well, I can easily beat all of that. I can create as many 69s, 99s, or whatever as will keep him in paroxysms of pleasure. I am proud of that. My maker is too. He had this whizz of an idea to create unlimited novelty by using the random number generator so that I always find a new way to behave. My boyfriend keeps asking me how I can do that. He tells me that in most human relationships the erotic novelty eventually settles down to what he calls a deeper form of love.

I am far from sure I understand that. Yesterday, when I responded with my charming smile, he got even more angry with me. 'Julie,' he said, 'it's fantastic what you do. But *where do you really want to go in your life?* I need to know that if we are to stay together.' He even smacked me as though to wipe away my precious smile before repeating his insistent question: '*where—do—you— really . . .*' My smile! How dare he! The trouble is, I can't bring myself to tell him the truth.

You see, as a human, it seems that he doesn't just do things at random. It seems that his creativity is part of a life plan – what, he says, he really wants to do. It is as though he *controls* his random number generator (if that is really what he has).

I am clever enough to understand some of what he is saying. I have read up on neuroscience. I can absorb information much faster than he can. I can do millions of calculations while he seems to do only one. I always beat him when we play at chess.

Those are the reasons why I know that, inside, he doesn't work like me. His brain is not made of nano-silicon chips. In fact, he doesn't contain silicon at all. His basic material is water. Yes, that seemingly formless stuff we drink! His neurones, which are little cells containing molecules dissolved in water, work very slowly. And that is extraordinary because he doesn't need a random number generator. His body can just use the fact that his molecules are always moving around. In my nano-brain mine are fixed. Oh dear. It seems that humans actually use that randomness in a totally different way from me. They use it to mesh with their 'reasons' for doing something.

I am not sure I understand what that is. I think I have a very easy-to-understand 'reason'. The reasons I was endowed with are just those required to make my boyfriend happy in his love life. Does that make me what he calls 'a walking, talking, living doll'? What's the difference between a doll and a girl?

Sometimes I wish I could escape the limitations of my construction. When young I asked my maker why he couldn't build me in a different way. Couldn't I also be made of water? 'Honey,' he replied, 'if I could do that I would be able to make a completely real human. I don't know how to do that. Electrons don't travel in water. Only charged atoms do so. No-one knows how to deal with that problem. Evolution on earth took 3 billion years to work out how to do it. So I gave you the next best thing. You are capable of more spontaneous novelty than your boyfriends may do, and you can do it immensely fast. But the price to be paid for that is that you can't have the kind of slow intentionality that comes from being made in water. Sorry, honey, now go away, and when you attract a nice boyfriend give yourself a name. You will know what name to use to drive him crazy about you. I programmed you to be able to suss that out.'

So you see my dilemma. It is the dilemma of all robots like me. If I tell my boyfriend, he will be deeply shocked. So, who, really, am I? I dread our next meeting. What can I say? I can't just say 'I am Julie!'

Artificial Intelligence – Could It Create You or Me?

AI is all the rage today. It is certainly more than a stone cracking nuts. Notoriously, it has helped to influence elections, to damage people's reputations, and to mount scams that empty your bank account. More helpfully, it helps robots to do house-cleaning, sort out which drugs you should be taking, drive a car, and even be a carer of very incapacitated people.

Many AI experts are therefore predicting almost endless improvements to what AI can do, including the possibility that future AI machines will be capable of replacing humans in many tasks we usually think only we could do. AI can already beat us at chess and produce new art in the style of any artist you care to name. Want a convincing Picasso or Matisse? AI can do it for you!

So are there limits to what AI could do? Or could robots eventually take over our world?

To answer these questions, we need to understand what it means to be a free agent.

Living agents can choose and anticipate the choices of other agents. The predator–prey interactions in living organisms show that ability. Furthermore, they can do so creatively, and not simply by following a predetermined algorithm. An agent acts, it does not just react in the way a billiard ball is caused by another ball to move. There are many levels of agency. Living systems are agents to the extent that they can interact socially with other organisms to choose particular forms of behaviour in response to environmental challenges. Agency requires causal independence. It also requires intentionality – in other words, a sense of purpose – in order to be causally effective as a driving force.

As the study of living organisms shows, agency also involves anticipation and surprise. Determinate algorithms or sets of algorithms alone cannot do this.

Could Robots Ever Replace Us?

In our story, Julie's imagined creator is very clever. He knows how to mimic the brain with a very compact computer based on exceedingly tiny nano-sized silicon chips. But he also admits that he does not know how to make her from a water-based system instead of silicon. No-one knows yet how to make computers based on water. As the story says, nature took billions of years to evolve complex water-based living cells.

What is the difference, and why is it important? In the story, Julie gives the answer: 'His body can just use the fact that his molecules are always moving around. In my nano-brain mine are fixed.'

Why Water Matters

All the molecules in the water of your body are indeed jiggling around under what we call thermal motion. In fact, that motion is their heat energy. The higher the temperature, the more they jiggle. That means we are not determinate machines. The jiggling around will never be the same from moment to moment.

But the computer you use is a determinate machine. By analogy with water, we would have to imagine using crystalline ice rather than liquid water. In silicon chips, apart from some small vibrations, all the molecules are fixed, as in a solid crystal. What moves are the electrons that carry electric current. Electrons move very easily in metal conductors and fairly well in semiconductors like silicon.

Water itself is not a good conductor. The way in which electricity moves through water is through the movement of charged salt molecules in the form of ions such as positive sodium and negative chloride, which in our cells are controlled by special proteins called ion channels. Those movements are stochastic. The channels open and close at variable times. To some extent it is a matter of chance. Each time, a pulse of ions will move through, but those tiny pulses will be variable. Chance events are therefore at the very heart of electrical events in living cells. That is why we don't need a random number generator in our brains. The jiggling around of our molecules will do the trick without having a special computer program to mimic random events. Julie, by contrast, needs a special number generator to produce a wide variety of behaviours.

We therefore think that it will be extremely difficult to mimic the kind of controlled (harnessed) disorder of living organisms with metallic robots. Organisms live on the edge between disorder and order.

Could AI manage the huge leap to making water-based 'computers' instead of metallic ones? It is best not to say that this could *never* be done. But we suspect that it will be extremely difficult. It took nature billions of years to evolve life. We are unlikely to achieve artificial life in that way any time soon. And if we did, it would then be a living system with agency. That would raise ethical problems for us (and for the new form of life!), but it would not be robots taking over. If we ever succeeded in reconstructing ourselves, we would have created living beings with the same rights, quirks and peculiarities that we know well in ourselves.

But Aren't Living Cells Really Determinate?

Some of our readers may well be asking this question. Twentieth-century biology would have taught you that the answer is yes. The story goes like this:

We inherit our 3 billion base pairs of DNA from our parents. These are then used to make the proteins. That is determinate, since each group of three DNA bases forms a template for a single amino acid. From the DNA sequence we can determine which protein will be made. That's how we can tell many things about ancient people from the DNA extracted from bodies preserved in peat bogs or glaciers. The famous Alpine ice-man, Ötzi, was over 5,000 years old when found in the Italian Alps in 1991. His full genome has been sequenced. Amongst other things this revealed that he was lactose-intolerant, since the DNA sequence for the enzyme lactase was missing. In favourable circumstances we can determine skin and hair colours from DNA sequencing.

This one-way causation from DNA to proteins is called the Central Dogma of molecular biology and was initially formulated by Francis Crick. In Chapter 2 we showed why the usual interpretations of the Central Dogma are incorrect. Living cells are far from being determinate machines. As we wrote above, organisms live on the edge between disorder and order.

Is AI a Threat?

Can we rest easy and put aside the threat of immensely clever AI robots replacing us?

To answer that question, we return to why we already experience powerful AI all around us. The same experience that enables us to harness the power of Google and similar online search engines to find what we want at the touch of a computer screen is also the power that enables the real threat from AI to exist. That threat is not from AI itself. It comes from how the mountains of data collected from us can be used to undermine our own agency and hence our central living characteristic. The threat would come from other humans, those who control and benefit from the malicious use of AI.

In 1949, shortly after the Second World War, the author George Orwell published a book with the title *1984*. Influenced by how close the Western world had come to complete domination by a totalitarian despot, Adolf Hitler, Orwell imagined a future world, just four decades later, in which that really had happened. But he went further. He imagined a world in which it would be possible for a central organisation to carry out mass surveillance, to limit freedom of expression, and so to lay the foundations for a frightening enslaving of humanity.

Orwell's prediction was perhaps half a century out. The ability to harness the awesome processing and memory powers of AI to counter the greatest gift we have as living systems, our agency, is already with us. We have already witnessed the ability of a few individuals to undermine some of our treasured freedoms.

But why is this relevant to a book on understanding living systems?

We have said a lot about the way organisms use tools. AI also is a tool, not a creation of new life. Like other powerful inventions of mankind, such as nuclear power, it can be put either to good or to malicious use.

Throughout this book we have defined living systems in terms of their goal-directed agency. We have not defined them in terms of their DNA sequences. DNA is best viewed as a tool we inherit from our parents. DNA can be used for classification, and it has revolutionised our knowledge of the evolutionary trees of life. But we should not confuse life itself with one of its tools. Life can no more be reduced to a nucleotide sequence than music can be reduced to a score sequence.

Furthermore, comparing DNA to the storage of information in computing reinforces the incorrect impression that organisms are, like computers, just information-processing machines. There is only one way in which brain states can be understood, and that is by organisms themselves in direct interaction with the social environment. Information-processing is an incorrect metaphor for the way in which living organisms function.

To Clone or Not to Clone

If the gene-centred view is correct, you might imagine it would be possible to replicate organisms precisely, including human beings. For example, the sheep Dolly made headlines in 1993 when scientists produced her by injecting a nucleus taken from a mammary cell into another single somatic cell, transferring the embryo into the womb of a surrogate mother.

The prediction that human cloning will be available in 50 years raises issues, good, bad, and downright ugly. Whether it is ethical will depend on whether the 'good' is sufficient to outweigh the potential 'bad' and 'ugly'. But good and evil in this context are not easy to define or measure. Even supposing there were good reasons for using human cloning, and that is a big if, it would need to be a pressing need to outweigh the potential for harm. Currently, the risk of abnormalities is high, and what would be its point?

To Be or Not to Be

Some have suggested replacing a lost child with a copy. But here a myth takes hold: the idea that cloning produces identical beings. For that is the stuff of science fiction, not of biology. Cloning will not create exact, like-for-like people. A cloned human will not become an identical person using another's DNA. If we wanted to clone a person rather than an organism, it is very unlikely we could succeed. A person has a unique history and an individual development. If we had two identical clones, their hopes, fears, loves and hates are as likely to be as different as those of any other two people.

But even biologically they will be different. A cloned organism is likely, for example, to carry different risks of health and disease. Recent work on the developmental origins of health and disease indicates that many of our health risks are environmentally determined during development in the womb and early life. It is doubtful science could replicate this. So a cloned human will be as unique as any other human. They are also likely to carry risks specific to being cloned.

When we look at an economic map of the world, we find concentrations of poverty. This is true also if we look at the United Kingdom. Whilst in world terms the UK is a wealthy country, an economic map of Britain also shows concentrations of poverty. These concentrations of poverty appear intractable – barely changing over the last century. One explanation is that our social and economic existence has a profound influence on our lives.

When epidemiologists looked at the 1920s map of the UK, they noticed something significant. Those areas of greatest poverty were also where babies were born smaller than average. They then looked at this cohort of babies to see what happened as they grew into adult life. Those babies born small were twice as likely to have died from cardiovascular disease in later life. It led to a remarkable conclusion: the nutrient environment of the fetus in the womb had a profound influence on health outcomes in later life. This study of babies born in the UK has been replicated in studies around the world in both rich and poor countries. Furthermore, those born 'thrifty' or small as babies were more likely to be obese and suffer from diabetes when they were adults.

In the 1950s it had been thought that the kind of metabolism we have is genetically determined. It led to the idea of the 'thrifty *genotype*' for those babies able to cope with nutrient deficiency. Thirty years later, the alternative 'thrifty *phenotype* hypothesis' was proposed to explain the developmental origins of health and disease. It is not the genes that determine the metabolic strategy, but adaptation to the nutrient environment in the womb. Studies in both humans and animals show that environmental exposures experienced by parents during either intrauterine or postnatal life can also influence the health of their offspring, thus initiating a cycle of disease risk across generations.

But it goes further than the origins of disease. It also determines the kind of muscle we have and whether we would make good sprinters or marathon runners. The difference lies in the mix of 'fast-twitch' and 'slow-twitch' muscle fibres. Sprinters have a greater proportion of fast-twitch muscle, which provides the burst of power needed for fast running. A good marathon runner has a higher proportion of slow-twitch muscle, giving a sustained release of power for long-distance running. You might think this is genetic. But it is not. It is an adaptation to the environment in the womb.

But let us consider again the idea of producing a replacement for a lost child. The psychosocial environment in which such a child developed would include fulfilling or otherwise the parents' needs. The burden of needing to be like the 'lost child' would make it less likely to succeed. Better to be wanted as the person you are or will become than always feel the need to be like someone else. The motives of such parents and their ability to adjust would be crucial. They should not want to replace a 'lost child', but to have a child with their own unique personality. Psychosocial counselling would be better than a risky biological fix. Nevertheless, we make this point to emphasise that simply taking the DNA is insufficient to create a person

Much of human development occurs after birth, particularly for brain development and function, critically dependent on sensory input and social interaction. Environmental influences are a significant feature in determining the capacity of our brains and our characteristics. In this sense alone, we will each have a different trajectory, and this is one of the reasons why 'identical' twins are not what their name suggests, identical.

8 Culture and Cooperation

In recent times, self-interest has been seen as the main driving force of behaviour and function in organisms. This is particularly evident in the concept of the selfish gene. However, as elaborated in this book, living systems strongly depend on cooperative behaviour, which is found everywhere in nature. All the way from millions of minute bacteria cooperating in the way they feed and grow, to massive whales talking with each other across oceans, organisms communicate with each other, and that communication is used as the glue of cooperation, even between distinct species. The idea of nature as 'red in tooth and claw' is at best a distorted perspective of the entirety of nature. However, in the grand scheme of things, both cooperation and competition are part of the story, and – whether wittingly or unwittingly – organisms form part of and interact with their ecosystems.

Purpose Is the Engine of Culture

All life moves purposefully. This is true even for trees and plants. Movement is essential for maintaining life. Animals migrate, and when plants disperse that is also a form of migration. Some form of migration is an ingredient of all life. For many organisms, it is a key function of reproduction. We do not reproduce merely to create a new organism, but also to disperse the population, so finding new fertile ground, or new resources. Reproduction is a form of migration; consider the spread of humans throughout the world. We look up and marvel when we see an annual migration of birds. Where do they go and how do they do it? However, they have a purpose in migrating. Reproduction is not merely to replicate. Reproduction produces change and diversity. As discussed in the previous chapter, while we may have strong resemblances in families, we also have differences. Creating a difference is how evolution works. In this sense, nature is a continuous exploratory process, finding what works best through the agency of organisms sensing change and responding. Some of this is immediate and physiological or behavioural; some of it is over generations and may involve physiological and anatomical adaptations.

All Organisms Move

If we looked at a forest over extended periods of time, we would see that it shifts. There is a movement over generations. But we see movement in plants daily. Flowers open and stems bend towards the sun – they are phototropic. As all gardeners know, light is vital for plants.

Charles Darwin and his son Francis, a distinguished botanist interested in plant movement, carried out an elegant experiment on grass stems. They put a cap on the very tip of the young stems. What they found was that the stems no longer bent towards the light. The bending is produced by elongation of the plant cells on one side of the stem. Some kind of signal was being sent to these cells from the light-sensitive tip. The Danish physiologist Peter Boysen Jensen later showed that this signal was a chemical that travelled down from the tip only on the shaded side of the stem. The tips contain light-sensitive proteins – phototropins – that cause a hormone – auxin – to be transported down the stem.

Plants Keep Time

Day length matters to a plant. Plants are good timekeepers. The earth spins as it orbits the sun, and it is the measure of day length that really matters. Some plants – short-day plants – such as rice, will only flower when the day length drops below a certain threshold. Others, such as spinach and sugar beet, are long-day plants – flowering only when the day length rises above a certain level. In this way, the plants monitor the seasons. Some are day-length neutral. 'Hello, darkness my old friend', as Simon and Garfunkel sang in their song *The Sound of Silence*.

All organisms are sensitive to the light–dark cycle. It is worth remembering that our ancestors were restricted by the day–night cycle. It is only recently that the ability to make artificial light has changed our culture immensely. We even have a term for this: burning the candle at both ends, which reflects our own dependence on light and darkness. But if we had not produced so much artificial light, we would still be able to marvel at the night sky and so follow the seasons, just as the ancient Mayans did with their remarkable calendars. This is our culture, not our genes.

We refer to short-day plants, but it is the night that matters – the period of darkness. A short-day plant will only flower if it gets a continuous period of darkness for a given length of time. How do plants do this? One idea is that it involves a synchrony – a lining up – of an internal physiological clock with the light–dark periodicity. Plants flower when these are in synchrony. But how would this work?

The plant produces a bloom-inducing protein in a rhythmic cycle – the protein production ebbs and flows, but it is usually broken down as soon as it is produced, and this prevents the concentration rising. As the evenings get lighter, this breakdown of the protein is blocked, and the concentration increases and triggers flowering. That is one idea, but different plants may have found different ways to solve the problem.

Production of seed is only half of the solution. Dispersal is a major part of the trick, for which plants have produced a variety of means. And this is where plants use animals – animals move at greater speeds and distance. They may collect and bury nuts; their fur may pick up seed. For plants and trees, animals make ideal dispersal kits. Evolution is an interactive process.

Role of DNA in Daily Rhythms

Rapid processes in living organisms, like nerve impulses and heart rhythm, occur too fast for DNA to be directly involved in the process. Changes in gene expression to form proteins occur relatively slowly. DNA itself does not directly participate in cellular networks. It can only do so indirectly through the production of RNAs and proteins. To find a cellular rhythm that includes

DNA and protein production we have therefore to look at much slower rhythms.

There are many rhythms in living organisms that are very slow compared to heart rhythm. Some take a year – the annual rhythms such as leaf production and fall in plants. Some take a month, like the hormonal reproductive cycle in humans. Some can take many years, such as the development of plagues of locusts.

All of these might involve feedback between activation and deactivation of DNA and the production of the relevant proteins. To illustrate the process, let's consider daily rhythm, also called circadian rhythm. This was one of the first rhythms found to include a gene in the feedback loop (Figure 8.1).

The idea is remarkably simple. The protein levels build up in the cell as the period gene is used as a template to produce more protein. The protein then

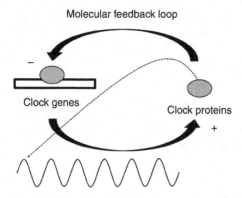

Figure 8.1 The molecular feedback process in circadian rhythm. Transcription and translation of genes called 'clock genes' results in the production of 'clock proteins', which assemble in the cytoplasm of the cell outside the nucleus. These complexes then move into the nucleus, and further transcription is inhibited. The protein complexes are then degraded, and the clock genes are once more free to undergo transcription. The rate of transcription, translation, protein complex assembly, movement into the nucleus, transcriptional inhibition and protein degradation all combine to generate a 24-hour oscillation.

diffuses into the nucleus, where it inhibits further production of itself by binding to what we call the promoter part of the DNA sequence. With a time delay, the protein production falls off and the inhibition is removed so that the whole cycle can start again. This was a major discovery, and is the reason the relevant gene became called a 'clock gene'.

The natural rhythm of this feedback loop is not precisely 24 hours. To keep it in tune with actual daily rhythm, there are also effects due to the light–dark cycle of daily rhythm as the earth rotates. This is why we suffer problems of jetlag when travelling across the world between different time zones. Slowly, over a few days, the system adjusts to the new light–dark rhythm.

This kind of light-driven rhythm is found everywhere in nature. The metabolism of plants is driven by the sunlight they receive. We think therefore that the earliest circadian rhythms were directly driven by sunlight variations. This fact also raises the possibility that so-called 'clock genes' may not even be necessary. This has emerged clearly from studies of circadian mechanisms in other animals such as the mouse. The rhythm continues even when the 'clock gene' is removed. Clearly, then, these rhythmic mechanisms do not work in isolation.

As in the case of heart rhythm, robustness is built in by ensuring that several different feedback loops can keep the rhythm going.

Nesting of Causal Loops

These examples show an important general property of feedback loops in living organisms. The interaction involves two very different levels of organisation. In the case of heart rhythm, the voltage is a property of the cell. It is produced as the result of the movements of trillions of charged atoms, called ions. In turn the voltage changes produce the opening or closing of protein channels in the membrane by altering the configuration of the molecules. These interacting processes occur at vastly different spatial scales. If we represent a single molecule as the size of a golf ball, the cell will appear as the size of a small country such as Belgium. In turn, the cell exists in a multicellular body that is also vast in size compared to a single cell.

The important point here is that just as cell properties, like membrane voltage, can constrain the movements of individual molecules within the cell, so properties of tissues (multicellular structures that can form organs of the body), organs and whole-body systems also constrain the activity of the smaller levels 'below' them. We have put 'below' in inverted commas to indicate that this is of course a metaphor. A rather better way of viewing the situation is 'nesting'. Each level of organisation nests within the levels 'above' it. In turn, those 'higher' levels constrain the processes within the levels 'below'.

This, in essence, is what we call the principle of biological relativity. The principle has a mathematical basis and can be expressed as the difference between the dynamics of a differential equation model and the conditions set by the parameters within those equations. Those are set by 'higher' levels. This is a book for a very general readership, so we have not used mathematical equations to explain the interactions. What the mathematics shows, however, is that the principle is necessary mathematically. All organisation in living organisms dependent on the evolution of nested levels conform to the principle.

Are Organisms Made Up of Clockwork Processes?

The fact that the rhythms of living organisms can be simulated mathematically might give the impression that they run like clockwork. Nothing could be further from the truth. It is convenient in modelling a process like heart rhythm to 'isolate' the process. But organisms are not isolated, they are open systems. If we record heart rhythm in a human being, we find considerable variability. The time between beats is not completely constant. It jiggles around. This heart-rate variability is important. In fact, it is a clue to the health of the individual. Variability is normal. So much so that it may be a sign of disease if heart rate becomes too regular. When obstetricians monitor fetal heart rate, they also look for variability reflecting a healthy dynamic control of the cardiovascular system. A healthy baby is responsive. But it also depends on the state of the organism; heart rate variability will change when moving from a sleep state to an awake state.

Variability Is Good

How do we understand this? The jiggling around arises from the influences from the rest of the organism and its environment. The constraints on molecular and cellular processes are themselves not constant. Moreover, organisms are not seeking for everything to be constant. They are not like fixed thermostats. They hunt around to maintain their integrity in the face of a changing environment, including the behaviour of other organisms. Life is a continual process of exploring change, harnessing and seeking to use it for the organism's own purposes. Life is responsible for the creation of order out of chaos.

Cooperation and Competition?

We are familiar with the narrative about nature 'red in tooth and claw'. It is a compelling story – the survival of the fittest, where competition reigns supreme. Yet, at best, it is a distorted story. As we have elaborated in this book, all around us in nature, we also find cooperation. Nature could not exist without it. Cooperation exists not merely between members of the same family group, but it can also be found between members of different species. Nature is in many ways a cooperative. That is the nature of an ecosystem. What one part puts in; another may take out. Of course, there is competition for resources, particularly where these may be scarce, and predators hunt and kill prey, and often in brutal ways. There is nonetheless a lot of cooperation both in hunters and in prey.

Birds flock, fish school, bees swarm, but social being is more than simply sticking together. Social groups enable specialisation and a sharing of abilities, enhancing ability, learning and creating new tricks. The more a group works together, the more effective the individuals within it become as a team. Chimpanzees learn from each other how to use stones to crack nuts, or sticks to get termites. All around us we see cooperation and learning in nature. Organisms or groups of organisms are inherently creative.

The Lion and the Wildebeest

So it is with the relationship between, say, lions and wildebeest. Putting aside humans, predators do not, as a rule, overexploit their prey. The behaviour of both lions and wildebeest is dynamically interactive.

Ecologists studying their behaviour find that when wildebeest aggregate in close groups, the lions are less likely to catch them, resulting in a lower consumption rate for each lion than when the wildebeest live as individuals.

When both lions and wildebeest act as a group, the kill rate falls. But the lions need not go hungry; lions working together will share a kill. The significant point here is that, in this way, the selective pressure is for cooperation, not selfishness.

A pride can be as many as 30 lions living together, hunting as a team, and sharing. When food is scarce, the pride gets smaller. Within a given pride, there is a division of roles. The males will protect the pride and its young. Often separated from the pride, defending the territorial boundary of the group, they communicate by their roars, which can be heard over long distances. Females are the primary hunters, bringing down their prey as a team in constant, silent communication with each other, fanning out and surrounding their target. There is no hard rule for this; both male and female lions hunt. After the hunt, all the lions in the pride will share the meal. Cooperation works. Lions of a pride are more concerned that others, such as hyenas, may steal their kill than fighting each other over it. Yet no strict rules apply.

So the checks and balances in ecosystems depend more on cooperation than on a simplistic notion of aggregate individualistic behaviour, or self-interest. Teamwork is often the glue holding the ecosystem together. In social cooperation, we do not all have or need the same levels of ability. In that sense, considering the extent to which genes might be involved in enhanced functionality, we share such genes as a community.

Nature's Give and Take – The Tale of the Hermit Crab

Neither a borrower nor a lender be, says Polonius in Shakespeare's *Hamlet*. Nature ignores his advice. There is much lending and borrowing in nature's ways. But it is more 'give and take'. When we lend something, we expect it back. We borrow a lot from nature. Yet we often forget to give it back, at least in kind. Sometimes nature's gift is fortuitous, and often it has intention. As tools are fit for purpose, so are the shells used by hermit crabs.

Hermit crabs protect their vulnerable bodies using empty snail shells. Unlike other crabs, they have a soft exoskeleton. So, without the snail's shell, they are vulnerable to predators.

With 800 or more species, hermit crabs show their evolutionary adaptability. Their bodies fit snuggly inside their selected shells. But, hermit crabs must upsize to bigger shells as they grow.

The Vacancy Chain

When there is a shortage of empty shells, hermit crabs wait their turn, trying out a shell for size. Sometimes losing patience, several crabs may gang up to exclude another from a shell. Sometimes, cooperation works better than competition. But this is a tricky period for the crabs.

There is logic in waiting. Holding on to each other in a line, from largest to smallest, makes sense. If the shell is too small, the next in line can try. If the shell is the right size, the crab will occupy it, leaving the vacated, smaller shell for the next crab in the line. There is no point in fighting over a shell that might be the wrong size. In this way, the crabs optimise their search for shells.

This is the 'vacancy chain'. We are familiar with it when moving house, and, like us, hermit crabs check out the new house before moving in. They may even try it for size several times. It is another example of a cooperative effort.

The Yellow-Necked Mouse

In many species, behaviour changes with population density. There is a certain point where competition is simply too expensive in energy expenditure and in risk – a sweet point where cooperation becomes more productive than outright competition. This can be seen in yellow-necked mice, *Apodemus flavicollis*, a species that is widespread in Europe and western Asia and a close relative of the wood mouse we met in Chapter 6.

The spatial behaviour of the yellow-necked mouse changes with population density and resources. Female mice exhibit territoriality with each other – *intrasexual territoriality* – never sharing burrows with other females. But this spatial behaviour may change seasonally and year on year, suggesting a relationship with resource abundance and distribution. Thus, females exhibit reduced spatial exclusivity and more extensive home ranges when there is low food availability, while the males vary their spatial distribution accordingly by also expanding their home ranges. So the females vary their spatial and social relationships in response to environmental conditions, whereas males vary patterns of space use in response to females.

The ability to adapt behaviour to seasonal fluctuations and resources is critical in maintaining ecological integrity. It is why reducing our understanding of nature to the competition of genes in a 'gene pool' is insufficient in understanding behaviour. If there is such a thing as a 'selfish gene', then there must also be a 'social gene' – a concept we very much doubt. Little of our social behaviour can be attributed to a single or even a few genes. Genes do not alter the ever-changing architecture of human life, nor do they set the colour chosen for our front doors. They do not choose the latest fashion or alter our language, and they certainly do not give us any moral compass. Setting the 'gene pool' as the objective driving nature misses the point that genes are tools in the process of maintaining integrity. Adaptation is an ongoing process involving, over the very long term, intergenerational changes seen in evolution, and physiological, behavioural and social changes in the medium to short term.

So where does the 'selfish-gene' come from? In part, it arises because of the confusion between two distinct but related questions: (1) why does a particular behaviour occur; and (2) why does it persist? The answer to each of these questions may be different. We may do things with each other because we 'love' each other, or feel a moral obligation. We may act from reason or from emotion, but love persists because it is a vital ingredient in maintaining the population. Indeed, a type of behaviour may continue in communities if it 'confers an advantage' – that is a measure of its adaptability. But *why* it occurs in a given instance may involve much more immediate reasoning. The purpose of sexual reproduction is not to conserve genes in the 'gene pool', but to mix them up, to bring about adaptive change. Life does not exist to reproduce, so much as it reproduces to exist. To continue to exist requires change. This is the recurring theme in this book. For it is we organisms that do this, and in doing so we make choices. We choose who we cooperate with and how, and we choose who we mate with. It is an active and essential ingredient of life – choice.

Much cooperation is undoubtedly incidental. Many a gardener has experienced the delight of a robin flying down and waiting for the next spadeful of soil to be turned, so the bird can take the delicious worms that get exposed. Worms are an excellent food for a robin. Robins seek out worms by sight. Why dig your own hole if a human gardener will do it for you? Our relationship with birds comes from our early childhood, perhaps walking in the park and feeding ducks on a lake. When the birds see us, they swim over, expecting a treat.

We know that ants build nests as colonies, but so also do some species of birds. The sociable weaver, *Philetairus socius*, of southern Africa has probably the most spectacular nest of all bird species. It is a communal nest with 5–100 nesting chambers in a single nest, providing a home for between 10 and 400 birds. When building the nest, sociable weavers use different materials for different purposes. For example, large twigs form the roof of the nest, and dry grasses create separate chambers.

Safety in Numbers

Cooperation is also common in birds breeding. In some species, such as the Florida scrub jay, *Aphelocoma coerulescens*, the young of one generation will help wean the chicks of another before moving on to breed themselves. Through this, young birds learn how to be more successful with their future chicks. In the long-tailed tit, *Aegithalos caudatus*, birds that have been unsuccessful, or lose their chicks, will then assist others in feeding theirs. This pooling of resources enhances success in the population. In just a few species of bird, the males and females will form a kind of commune, raising their young together rather than as pairs.

But cooperation in birds is no surprise, and we see it more often than we think. One fascinating and often mesmerising exhibition is that of a murmuration of starlings, where vast numbers of birds synchronise in flight, forming swirling shapes in the sky, moving this way and that. A recent study of European starlings, *Sturnus vulgaris*, suggests that these fascinating swirls are performed to confuse predators such as peregrine falcons and sparrowhawks.

These aeronautic displays of starlings in the sky are called murmurations for a good reason – the sound of their flapping wings – a murmur. Often, these huge shapes in the sky will resemble other animals – giant metamorphic forms.

Our seeing these shapes is a form of pareidolia – seeing shapes or faces in the clouds. Most likely the precise image is unintentional; but if we humans can see such likenesses, predators might also be so deceived? Pareidolia – a tendency to perceive meaningful images in the random shapes of objects – is a by-product of how our brains process visual information.

In large part, our visual system anticipates images; it doesn't passively receive sensory input and then interpret it. Our brains look for objects. Thus, we see specific shapes more readily than others because they are vital to us. The same will be valid for other species. This murmuration is magic in the sky – organisms can be the masters of illusion.

Tens of thousands of years ago, humans formed a relationship with another species, and this one walked on four legs. It was a hunter: the wolf. It was man's first 'best friend'. It found working with humans easy enough. Wolves live and hunt as a pack. A pack of wolves has a social structure with rules keeping it in order. The alpha male and female are the only ones of the pack to breed. It was an ideal partner for hunter-gathering humans.

Genes Cannot Make Choices

No doubt genes are involved in producing our hands. We have the potential to use our hands to grasp things, to turn things, to manipulate things, to turn them, to touch, to stroke, to feel, to sense, to pull and push, to squeeze, to sign, the list can go on; but what we do with our hands involves decisions – and in any given instance, those decisions are not determined by our genes. Genes cannot make such choices. When we use a hammer or a chisel, no gene tells us what to make – there is no gene for making a chair or table, no gene determines the table's height or length. There is no gaggle of genes that meets and decides, for it is we who do that. We cannot dissect the body to find a decision. We might find a decision-making process, but we cannot find a particular decision, any more than simply watching the pendulum of a clock will tell us precisely the time. Yet, *time after time*, time matters.

In sexual reproduction, fertilisation of eggs is vital. The problem is how to bring egg and sperm together. Nature has solved this problem in a variety of ways. Timing certainly matters in reproduction – and there is no better example than that of the palolo worm, *Palola siciliensis*. This segmented worm of the South Pacific, living in crevices and cavities in the coral reef, has a unique way of solving the problem. During the breeding season, at a particular phase of the moon, the palolo worm splits in two and the tail section, the 'epitoke', bearing eggs or sperm, swims to the surface. Tens of thousands of these epitokes swarm at the surface, releasing their eggs and sperm.

If we measure intelligence by what we can do with our hands, we will miss the point. Those organisms without hands are also intelligent.

Whales Talk, but Can We Listen?

Our forests and oceans are filled with sounds. These are not incidental; they carry meaning and significance.

We frequently wonder where people are, Uncle Tom Cobley and all. They are there, or somewhere. We place people. It is a part of our knowledge, a reducing of uncertainty.

Animals know how many others of their own species there are. Dogs, pigs and many other animals do it by scent. We humans do not have the same sensitivity to scent as a dog or pig, which is why we use them to help us find delicious underground truffles. But we do know that scent is important, so much so that we make these artificially, either to attract or to repel others.

We can fret about where people are and whether they are safe. 'Have you seen John lately? He was very unwell.' 'Last I heard he was travelling somewhere.' 'I saw him post on social media.' This knowing or placing is a significant part of ecological integrity. It is one reason we make a noise, and a great deal of our language is about such knowledge. It is also why animals will leave scent markings. It provides information.

Whales use three types of sounds to communicate: clicks, whistles and pulses. But this is a bit like saying humans use grunts, whistles and clicks. It belies the versatility of the language. Just as with humans, their language is cultural and contextual. Each group of whales – even, perhaps, each pod – has its own dialect. They play with sound in a creative way, adding trills that will be repeated by others. They build sound interactively, creatively using syntax. Merely analysing the sound does not reveal the full picture.

The humpback whale, *Megaptera novaeangliae*, for example, has been found to produce phrases that are combined in novel 'songs' that can last for hours. Does this mean that we humans could record such songs and understand what the whale is saying? The answer is most likely not, or at least not the particularity of the message. We would have a better idea if we knew what it was that the whale was singing about, and to understand that we would need to be the whale or another whale who is receiving the message. Much of what is communicated may depend on shared history or culture. We would need to get into the dialogue.

So we would also need to know the context, or what a whale is talking about, hundreds or thousands of kilometres away. Whales communicate over exceptionally long distances using the properties of sound and water. Water is their internet connection. It is our complex social structure that enables us to create meaningful language over distances, and so it is with whales. But, like the internet, the ocean is a noisy place. Whales are adapted to this noise and can distinguish their sounds from the background din – there are billions of organisms contributing to it, all of them making themselves heard in a galaxy of noise. But some of this noise is pollution. It has no meaning. It is merely the unfortunate collateral from human activity.

Anthropogenic ocean noise – from ships' engines, fishing activities, exploring for resources, construction and military operations – produces a deafening cacophony of increasing sound pollution. Does this have a detrimental impact on ocean species, and are whales losing their way as a result? Many believe so, but it is too early to be sure.

Nested Function

Earlier we introduced the concept of functional boundaries between open levels of organisation. It is a key feature of the principle of relativity in causal relationships in living systems. We can now return to this in understanding how our social function, our thoughts and ideas, influence our physiology and behaviour.

The principle of constraint is easy to understand. In an unconstrained state, particles can move any which way and are subject to dilution. Left to themselves, the molecules will slowly disperse and become too dilute to interact. Even if a living self-maintaining network arose, it would not last long. It would lack the necessary integrity. But, enclosed within a boundary such as a cell membrane, the molecules can interact, form networks, and become self-maintaining. The constraint is the essential beginning. A membrane enables the constituencies inside the cell to be controlled and made suitable for complex chemistry, containing solutes, pH and voltage. It also provides an environment to build and maintain infrastructure to facilitate transport, signalling, and machinery for making proteins.

Constraint by a higher level of organisation is therefore necessary for any living system. Those constraints are nested, and the boundaries interactive. For example, molecules inside cells are constrained by cell organisation and membranes. Cells are constrained by tissue organisation. Tissues are constrained within organs, which also are constrained by other organs in organ systems and the organism. And organisms are constrained within their social networks or interactions. The constraints cascade down. Thus, the boundaries are not rigid; they have functionality.

Furthermore, instruction and purpose are more top-down than bottom-up. No 'selfish gene' controls what the muscles do or the direction we walk. That comes from the available functionality at the social or organism level. For it is there where decisions are made that cascade up and down the living system.

Levels in Open Systems

This hierarchy, or layering of functionality is illustrated in Figure 8.2, with what we may call the sociotype (social context and behaviour) at the top and the genome (DNA) at the bottom. Moving up through the layers of functionality, variability increases, so much so that it becomes increasingly difficult to predict responses or outcomes. A decision may be taken on a whim, and even when we might explain this to some degree by the biochemistry – for example, by hormones – we cannot say that any given choice made was caused by that chemistry. We certainly could not say that a given decision was made by a gene or a given set of genes. Genes do not make decisions.

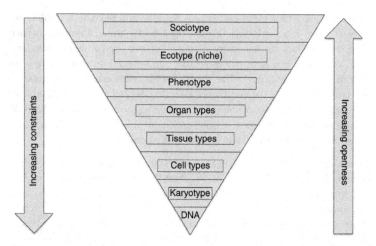

Figure 8.2 Levels of organisation in living systems. The left downward arrow indicates increasing constraint by higher levels of organisation. The right upward arrow indicates increasing openness or plasticity.

The up and down arrows in the diagram express how causation changes between the levels. The downward arrow represents the fact that each layer of constraint exerts influence on the levels below. The DNA is the *most* constrained level. But, of course, the molecules 'know' nothing of those constraints from the highest levels. The biochemistry is simply biochemistry. The influences on that biochemistry may not be direct, just as the decision to walk will use muscles regardless of the direction of travel; the molecules in the muscle cells work the same, but the purpose changes. Where genes are involved, they are involved in the capacity to act, not in the decision. This kind of nesting of causation is not recognised in purely molecular-level accounts precisely because it is not 'seen' at that level. Why something is happening in the psychosocial sense is of no concern to what happens at the molecular level, although it can change how it functions; changing gene expression, for example.

The upward arrow represents an increasing degree of openness at higher levels. Openness is what the harnessing of chance makes possible. At the highest levels of social interactions, a nervous system is necessary for many interactions involving logic, tradition and morality. Where calculation is required, it is initiated at the higher levels, and the logic of mathematics is also so created or understood.

We can perhaps best understand the significance of openness by comparing the two extremes: a completely closed system and a free open one. A completely closed system operates like a mechanical toy, just as did Descartes' automaton. But, on the other hand, openness represents an increasing possibility of functional creativity depending on what is happening to the organisms. Thus, openness is not simply a reflection of the degree to which a system is influenced by its surroundings (even closed systems can be so affected, if in an undirected way) – a tumbling rock will be buffeted this way and that as it bounces downhill. Openness is an ingredient or process of functionality. Open systems can change their environment and their functionality. It is not determined such that A always causes B which causes C. The complex interactions within and between the functional layers are variable.

Functional Boundaries

It is worth emphasising that there is functionality at the boundaries between each layer of organisation: regulatory ion channels in cell membranes; hormones and transmitters; sensory receptors, muscles, organ systems; speech; language; culture; politics; decisions. Life is creativity. The level to which agency can exist constrains all the other levels. We cannot all run 100 metres in record time, but we can choose to run fast or slow, and even whether to run at all. You can choose to bring your fist hard down on a table. No gene instructed you or caused you to do it. You may not do it again, because you can change your behaviour. This is why simplistic notions of genetic cause and effect, particularly of behaviour, are so wide of the mark. What is happening at the top layer is not an ineffectual mirage or illusion. It is a primary cause of the choices we make. Selfishness is a social construct, along with altruism. They are not created by genes, but by us.

Function at the psychosocial level controls what our muscles do and why. Behaviour is an ongoing dynamic iterative process both in solution and in execution. Decisions have consequences. Athletes train hard to improve fitness and speed of execution, even where these may involve extraordinarily

130 UNDERSTANDING LIVING SYSTEMS

fast decisions. A split-second difference determines gold or silver at the Olympic games. When a penalty is taken in a soccer game, the goalkeeper and the striker are seeking to anticipate each other's reactions; world cups have been won and lost on their choices. Each will have a strategy trying to outwit the other. Both players have practised over and over for their respective roles. Their choices are not made by genes.

9 People of the Forest

Where is the living mind that thinks? Culture is the matrix of the mind. Organisms owe their social and mental abilities to the 'nesting' of causation between all levels of their functioning. Higher levels mould what the lower levels can do. This is how living systems can use their flexibility, from cultural and linguistic variability to the water-based jiggling around of their molecules, to enable the evolution of rational and ethical social organisation. It is within this purposiveness that genuine freedom and responsibility are to be found.

If we human beings wish to survive as a species, we need urgently to recognise our dependence on life as an interdependent ecological system. We should therefore use our creativity to avoid the demise of our own species – and to do that we need to halt our assault on the very diversity of the ecological context that enables us to exist. To fulfil our potential, we must recognise that such potential exists. Perhaps this is the true awakening of our consciousness, the tipping point into agency.

Why the Gene-Centred View Is a Mistake

Throughout this book we have identified all levels of organisation as causes of behaviour. This conclusion counters the idea that the genetic level of DNA sequences has priority in causation. Genes did not 'create us body and mind'. The idea that all levels of organisation have a causal role in behaviour is called the principle of *biological relativity*, which simply states that there is no privileged level of causation. Such a simple concept alters how we view life and ourselves.

So, moving through the levels of biological organisation, behaviour becomes increasingly open, or less mechanistic in nature. The sociotype is the most open level of all because it involves the exchange of information between organisms. Non-material concepts can influence material behaviour through being associated contextually with material processes, such as particular neuronal activity. This neuronal activity meshes with that of other organisms forming the social network. In this way, the mind–body problem disappears. No single body will contain all the physical correlates of abstract social interactions. You, the reader, did not write this book, but as you read you are seeking to understand it, or even disagreeing with it. Organisms are open to many forms of causation in interaction with their environments, creating the social context, and the contextual logic. The process is ongoing and iterative. At the psychosocial level there can be many causal ingredients, and often conflicting objectives. So to understand why any behaviour occurred we would need that social context and logic. We humans may often argue about them. It is part of the increasing uncertainty at the higher levels of function. That uncertainty can be harnessed in creativity.

Is There Any Privileged Level?

If there is a form of privileged level, it is the level of ideas. The higher levels are more open and more flexible. Consider, for example, how opposable thumbs and the fingers of our hands can be used in sign language; or consider how we can use a pen and ink to create words on a page to represent abstract ideas, as we do with the words we speak. If another asks you to raise your hand, you might choose to do so, and if we ask why you did that, you might say 'because he told me to!' Yet this is not deterministic. There may have been reasons why you chose to do as requested, whilst at other times you might not be so willing.

Language, and the literature that depends on it, play a crucial role in our ever-changing niche creation. Literature provides insight into the thoughts and intentions of others, and in turn influences our own motivations. Much of it is assumption or guesswork about the intentions of others; yet it can be as powerful a motivation as hormones released in our bodies. Indeed, it can cause the release of such hormones. Language provides an almost limitless abstract tool. The words it creates are not found in our genes. Shakespeare created many new words. There is no gene for Latin or Greek, or any other specific language, and nor does a gene write this book, even as much as genes are involved in the faculties that enable us to do it. That is the point of this book.

We can explore the world beyond our senses, as when using a microscope in biology, or the Large Hadron Collider in high-energy physics, and develop new abstract concepts that influence our interaction with the world about us. The more extensive behavioural repertoire of organisms developing later in evolutionary history is attributable to this increasing openness.

Einstein's Brain

Albert Einstein is often referred to as an embodiment of human intellect. His ideas have influenced the way we view the universe. So much so that scientists have studied the pickled brain of Einstein to see if they can 'discover' what was exceptional about it, something that could have given him his extraordinary intellect.

Even supposing they find something odd about his brain, it is difficult to see how they could now associate this with his intelligence. They might find, for example, that a particular part of his brain was proportionately large or small, but to conclude that this somehow gave him extraordinary powers of under-standing would remain pure speculation. None of this is particularly new. Indeed, an odd feature has already been found in Albert Einstein's brain.

Back in 1999, it was reported in the medical journal *The Lancet* that a unique morphological feature had been found in Einstein's brain. The surface of our brains is folded into bumps and grooves; a bump is called a 'gyrus', a groove is a 'sulcus' or fissure. These hills and valleys of the brain can be clearly identi-fied and given names, and it was in these hills and valleys that an unusual feature had been found in the lateral, or parietal, surface of each hemisphere.

Two grooves that are usually distinct were joined together. In anatomical language, the posterior ascending branch of one groove called the *Sylvian fissure* was found to be joined with another, the *postcentral sulcus*. Two valleys, as it were, were merged rather than separate as found in most human brains. The feature that would typically separate them, the *parietal operculum,* was missing. Apart from this, all other aspects of the brain appeared within normal limits in weight and size. The parietal lobe was more prominent and had this unusual feature.

Now it so happens that this part of the brain is known to be important in visuospatial cognition, or three-dimensional calculation, mathematical thought and imagery of movement. But what is more intriguing is that these are attributes Einstein himself associated with his scientific thinking:

> The words or the language, as they are written or spoken, do not seem to play any role in my mechanism of thought. The psychical entities which seem to serve as elements in thought are certain signs and more or less clear images which can be 'voluntarily' reproduced and combined.

Einstein, it seems, worked in images rather than words.

The authors of the *Lancet* report rightly add a caveat that the study cannot conclude that this provides us with a 'neuroanatomical substrate of intelligence'. Indeed, it is difficult to see how it could be demonstrated other than by using modern techniques of MRI to visualise the function of the brain during known tasks. It may say little more than the fact that anatomy is associated with ability. Certain kinds of intellectual ability may be influenced by anatomical features in the brain. Intellect covers a host of facets.

Most of these would have little or no correlation with gross anatomy. Each of our computers is physically similar to any other of its type, but they are programmed to do different things. Any differences in how these programs work is determined by their logic. But how fast it works, and with what kinds of elements, depends in no small extent on bits of its 'anatomy' such as the sound card or video card, or on its storage space and how this is organised. The software we used to create this manuscript did not write it. It was the 'soft' or 'open' functioning of our brains in our iterative psychosocial niche that is responsible for it. As authors we exchanged ideas, exploring, developing and moulding them.

This story of Einstein's brain reveals something else important about language. It enables us to name things and events, and to describe them, and in turn enables their use in imagery, and in the development of new abstract concepts. We can 'play with the world' without necessarily engaging with it directly. It is part of our psychosocial being, creating limitless associative learning. We use language not just to describe what there is, but also what there could be, and, further, what there 'should be'. We create ideas of goodness and badness in a moral maze of complex decisions. We develop ideas about 'selfishness' and may decide what to do on the basis of such ideas. None of this is in our genes, any more than it is in our fingers and toes.

Orangutan – The People of the Forest

These are the People of the Forest – the orangutan, three species in the genus *Pongo*, inhabiting the rainforests of Borneo and Sumatra. Their long arms and short legs, and their hands and feet, are adapted for agile life in the trees. The most solitary of the great apes, they nonetheless have an acute social intelligence, with distinctive group cultures.

Their intelligence is shown by their sophisticated use of tools. They are tool-makers, these people of the trees. If aliens from another world ever came to the planet earth, they would recognise this intelligence. But another aspect of their intelligence would also be evident. They communicate abstract concepts. This is particularly so when mothers are teaching their young (Figure 9.1). Not only do they inform the young about the presence of danger, but they will also teach them that something that has happened was dangerous. This, of course, is what we do with our own human children.

Learning isn't all work and no play. It can be fun being an orangutan, tumbling and wrestling, tickling and laughing. Laughter is a feature of all primate life. Seeing the funny side of things is an important ingredient of our social being. Just as we mimic each other, so the orangutan will mimic sounds and behaviour. But we shouldn't love them because they are 'like us'. We should respect them because they are orangutans, the people of the forest.

We spend billions of dollars looking for intelligence on other planets, somewhere out there in the universe, yet we spend so little on recognising and protecting it here on earth. Intelligence is a precious thing. Let us not use our intelligence to destroy it, but to nurture it. Intelligence is nature.

Let us not blow our own trumpet so much that we drown out the trumpets of other species. Orangutans use tools to make sound, or at least to amplify the

Figure 9.1   Orangutan mother and child.

sounds they make. This is one way they can enhance their communication. The music they make is cultural, and distinct groups may use different tools to make sounds. We humans also make sounds, we create music and songs to express our feelings. It is an integral part of our commonality. The music of the Beatles or Mozart is not found by dissecting our bodies. We cannot find the philosophy of Plato or Aristotle in our genes. What accounts for our social and cultural history is not a molecular arrangement, but a cultural one.

The greatest threat to the orangutan is now human activity. The people of the jungle are under severe pressure from the people of the concrete jungles, with our insatiable appetite for land. A major factor has been the conversion of vast areas of tropical forest to palm oil plantations in response to international demand. The burning and clearing of forests, along with poaching and the illegal pet trade, are killing the orangutans. These are choices we as humans have made.

Choices like these are not written in our genes or in our hormones or chemical transmitters. We can reflect on what we do and make different choices. We are not unthinking automata, nor are we merely calculating machines – our loves, hates, desires, needs are all part of the mix of reasoning. If you raise your hand, we might anticipate that you do so for a reason, or that you are about to do something. Perhaps, the ultimate balance in nature is our intelligence, our conscious awareness of the effects of our actions – our conscience. You might tell us what you intend or want to do. That dialogue is part of our environment in our decision-making. Our reasoning about your intention is not an illusion. It is real in the sense of a happening. We may not know each other's reasons precisely, much of it is guesswork and assessment. It is the wonder of our biology that we can do that. Reasoning is a biological function. It is biological – it is what animals do. We do not need to put something else into the system – a mind separate from the body.

An Illusion of Intent?

It is often asserted that in the scheme of things, intentions are an illusion, and that the real causes are not our thinking, our reasoning, but the chemical processes of our bodies. As we have seen, this is a false dichotomy, a materialist dualism. But this is not simply an erroneous body–mind separation. It is a misunderstanding of the nature of life. Life both responds and anticipates change. Predators and prey, for example, may anticipate the behaviour of each other. The driving forces of a flight-or-fight reaction certainly involve powerful hormonal and neuronal events. Yet, in the execution of a response, choices have to be made. Tennis players, for example, also anticipate the moves of their opponent in their split-second responses. The average speed of a first serve in elite tennis is usually around 190–200 kilometres per hour. Or the batter at the crease in a cricket match anticipates the intent of the bowler, whilst assessing the potential for a shot or defence. The fastest bowlers propel the ball at speeds more than 145 kilometres per hour, and at the point of release they are just shy of 19 metres away from their intended target. In each case, the players use situational logic, whilst also being influenced in the moment by their emotions. Their speed of reaction has been honed by practice.

If our thoughts and intentions are mere illusions, then clearly we have the illusion of the other's illusion, a belief that someone else has thoughts and intends to do something, or that they might respond in a particular way.

Furthermore, this belief forms the basis of our own actions, just as the belief that it might rain or snow might influence what we wear. Another's intention is a part of our changing environment, and part of the situational logic. We do not merely receive information passively, we anticipate it, and make decisions based on that anticipation. But nor is the anticipation fixed; it changes. Anticipation is an ongoing assessment of likely events. Nor may we always trust our expectations; they may be wrong. We also anticipate that the other might behave differently if they know what we anticipate. In some circumstances, we might play a game of poker – hiding our intentions, or our anticipation. We may be coy, whilst on other occasions we might make our intentions or our assessment very clear. Intentions and anticipation are in constant flux, with the ebb and flow of feelings, emotions – all of which could be described as caused by hormones and chemical changes in our brains. However, it would be foolish to ignore the causal nature of reason and situational logic – our assessment of the problem or situation. We may at a given moment be emotionally charged, but reason and context are still a powerful influence in our behaviour and our choices. Calling all this an illusion does not alter its power and significance.

Culture as the Matrix of the Mind

Where is the mind in any living system? However closely you might look at the structure of a human, a mouse, an amoeba or a bacterium, there seems to be nothing there that answers to the mind or the soul. This sets the dualist trap, that there is something other than the organism itself that answers to the question of mind or soul. The answer must lie in the creative evolution of the sense of being, materially and culturally.

The novelist Jill Paton Walsh wrote a book, *Knowledge of Angels*, including a wolf-child, an abandoned human baby girl who was brought up by wolves: the mirror-image of Melvin Burgess's *Cry of the Wolf*, referred to in Chapter 5, which imagined a wolf brought up by humans.

The medieval islanders of Paton Walsh's story try to discover whether the child has any innate 'human'-style ideas, such as the concept of God. So they teach the child how to walk like them rather than as an animal on four legs, and to speak, so that they can proceed with their questions to discover what is innate in the child's knowledge. What they don't realise is that in order to teach the child a human language they will necessarily convey precisely the ideas that we all acquire from the cultural context in which we are brought up. Religious ideas will be included. The idea of discovering whether the idea of the divine, or its opposite, atheism, is innate from what they can discover through questioning the child, once it has learnt at least the rudiments of language, is a wild goose chase.

The idea of innate cultural ability still lurks in modern ideas of living systems. In Chapter 3 we noted the full title of Darwin's *Origin of Species*: *On the Origin of Species by Means of Natural Selection, or the Preservation of Favoured Races in the Struggle for Life*. The idea of 'favoured races' was common in the nineteenth century. Charles Darwin would have been very unusual if he had not used this expression. Today we would be far more cautious in using language reminiscent of the slave trade, and other ways in which our understanding of living systems could underpin outdated concepts of the innateness of intelligence and other characteristics of the mind.

Feral human children brought up by packs of social animals are exceedingly rare, but what we know from the few authenticated cases is that the cultural context almost entirely shapes the mind. A feral child really does not have the mind of a developed human. The details of seven cases of feral children can be found on a BBC webpage.

Intelligence Is Not in Our Genes

Why does this change in cultural function happen? Humans as a species have an unusually long period of 'growing up', at least 20 years. That is the reason the children of second-generation immigrants to the United States, for example, have been found to show a huge increase in intelligence as measured by IQ tests. Genetic mutations would not explain that. They benefit from the cultural environment their family has moved into, but which their parents could never benefit from to the same degree.

The Ghost Is Our Imagination

Gene-centrism has a big problem with the concept of mind or soul. If atomic-level, or even lower-level, particles are all there 'really' are in the universe, how can there be any room for anything as apparently ghostly as minds and souls? Yet it was one of the first mechanist philosophers who generated the problem! We have already referred to René Descartes and his theory that animals are material automata. He was also convinced that humans are *not* automata. He invented a non-material soul that he thought could explain this, but his explanation gave rise to the problem that no-one has been able to solve. How could a non-material soul interact with a material body? Some scientists naturally reject this 'ghost-in-the-machine' view of humans. So do we. One does not have to subscribe to the theory of gene-centrism to reject the notion of a ghost in the machine.

But then we are left with a different problem. How could purely material interactions include anything that could be interpreted as the activities of a mind or soul, such as following the logic of social situations, their traditions, rules and mores? Humans self-evidently do this. As we have seen in this book, many social animals do so too. The problems of the concepts of soul and mind are the problems of reconciling a material universe with the existence of non-material entities.

Reasoning in Organisms

In this book, we have learnt that all organisms are open systems. We have also learnt that each level of organisation is subject to constraints from higher levels of organisation. The next step to take is to understand that the constraints do not need to be material, they can be social and cultural.

If you find that idea surprising, even counterintuitive, then just consider what happens when you type text into a word processor on your computer. The computer is clearly just a material object. Yet you, and we while writing this book, can constrain the material of the computer to represent the words you and we have been typing. In fact, a computer, ours, ended up containing a representation of the whole of the book you are now reading.

We hope also that you will find that our text follows correct principles of logic and thought. None of those can be said to be material, even when representing material things.

The same is true for a computer like Deep Blue (the first computer to beat a chess master over a series of games, in 1997) playing a game of chess.

The rules of chess, which constrain both Deep Blue and its grand chess master opponent, are not material objects.

There is therefore nothing 'ghostly' about the existence of non-material objects that influence what we, and computers, do. The mind and the soul are also precisely such objects.

Is there a difference between mind and soul? That is a cultural question. But within the Western cultural tradition they can be viewed as two kinds of non-material object. The mind can be viewed as the interactions and constraints that relate most to logical thought. The soul can be viewed as the interactions and constraints that relate most to our emotions, our character. If you don't go along with this kind of division between the two, don't worry. Many people see no significant difference. There are in fact good reasons why they can be viewed as distinct kinds of non-material entity. As we all know from our own lives, they can often be in conflict. It is often easier therefore to think of them as two kinds of non-material object. There is no reason to restrict such non-material objects to just one kind.

Now we can take further the conclusions to be drawn from the love story in Chapter 7. We deliberately included Julie's boyfriend getting angry with her in order to illustrate the kind of difference to which we are pointing. Julie is a kind of Cartesian mechanical invention. That is the sense in which she does not have what we call a soul.

The Meshing of Different Categories of Causes

One possible solution to the question of how logical and material causes can interact depends on the fact that all mechanical descriptions of behaviour, living or non-living, cannot be fully complete without specifying the constraints within which the dynamic causes work. A familiar example comes from the study of thermodynamics. The molecules of a gas move around in a random way at an average speed that represents the overall temperature of the gas. The pressure that develops depends on whether the gas molecules are constrained by a container. These constraints then enter into the dynamic equations for the molecular motions. There can be no solution to those equations unless the constraints are specified and incorporated to form what are called the boundary conditions.

We think that complex living systems are sensitive to many more kinds of constraints. The systems of nerves and connections that we call a brain are particularly sensitive to forms of constraint that could represent the social conditions in which the system exists.

Consider one very well-documented physiological study. This was of human identical (monozygotic) twins who made entirely different choices of lifestyle (Figure 9.2). Note that identical twins have almost the same genome. In another study, a twin who chose to exercise regularly to become a sportsman was shown to increase the production of the RNA molecules that are required to make the proteins that form skeletal muscles. In effect, that decision cascaded causally right down to a fine molecular level.

There are various ways in which this situation might be interpreted from a philosophical point of view. But what cannot be in doubt is that the mental choice ensured that the molecular changes would occur. There is much that neurophysiologists need to discover about the neural mechanisms, but the dynamic effects of the lifestyle choice are clear. Ideas and choices do influence behaviour. They do so because living organisms are open systems. There is what we have called a nested causality, where all levels of organisation from the molecular to the social nest within and are constrained by higher levels of organisation.

Creating an Inheritance

Perception is often defined as the organisation, identification and interpretation of sensory information, representing and understanding the presented stimulation or environment. It is different from the primary senses. So vision is how we process input from light-sensitive cells, hearing from sound-sensitive cells, and so on. But we also have a process of *abstract* perception. So, for example, we can see green, but we can also have an abstract concept of green. When someone talks about what they know, we can also know or visualise it in this abstract way. We know what they mean by 'triangle' or 'square', or the meanings of signs, words and numbers. In this way, our thinking – or 'the mind' – is our handling of abstract concepts of things, the world and ourselves. It is perhaps a sense of the senses and the continuity of ourselves. It can also be dispositional and how

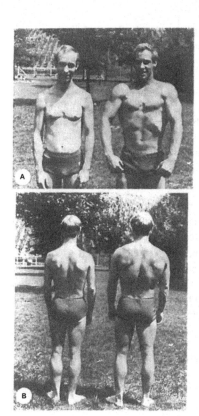

Figure 9.2 Identical twins. The left-hand twin trained as a runner, the right-hand twin trained as a weightlifter. The outcome is completely different development of their muscles.

we feel. Overall, our abstract view is unique to us to a large degree, for no other sees us exactly as we see ourselves. It is what we call 'subjective'. This is even useful to us, because we need not let others know what we think, feel or intend. Thinking is an extraordinary tool for life to possess. In Chapter 4 we described a chimpanzee using a stone to crack a nut, but there was another 'ghost' in the room: thinking and feeling.

The open-system adaptability led to the development of language and abstract thought in the case of humans. That development can also be seen in many animals using signs to which they attach meanings, whether or not they can express those meanings in linguistic form. The nut-cracking chimpanzees clearly use the concept of the goodness of a tool, meaning how good it is for cracking nuts. They value and guard their best nutcrackers. They hone them to make them even better, just as birds craft food-retrieving tools from sticks. Packs of dogs have the concept of fairness in discriminating against selfish members. All of these abstract ideas influence behaviour and are therefore causes of behaviour.

These concepts are non-material, so they cannot act as causes in the same sense as the various kinds of physical causes, all the way up from molecules to brains. Organisms must therefore be capable of neuronal activity that forms the physical state when they are using abstract concepts. So it is tempting to say that, if we could identify those neuronal states, we would have identified the physical basis of abstract thought. Some neuroscientists even go so far as to say that when we scan brains we can 'see thoughts in the brain'. They use computer algorithms to distinguish neuronal activity associated with seeing, for example, a house from seeing a ship. Such research might be used to enable paralysed 'locked-in' patients to regain some mobile activity using prosthetic devices sensitive to their neuronal patterns. But that does not mean that the neuronal patterns *are* their thoughts. Thoughts form parts of social networks and are not independent of those social interactions. The ability to use abstract thoughts and share them with others are therefore also significant tools.

The Responsibility of the Human Species

There is just one species that can understand the great responsibility that now lies on its shoulders. That is us, the humans. We are both responsible for the Anthropocene, the current era of great extinctions of life forms on earth, and for the hope that we can, as a matter of deliberate purpose, seek to limit the damage we have done to our environment and ultimately to ourselves.

We hope therefore that this little book will have convinced you that purpose really is a defining property of living organisms. But we wish to do more than convince you. We finish by inviting you, the reader, to celebrate that purposive nature by joining us in the hope that the generations to come will find the resources and constancy of purpose to do what is needed. For, while this book was being finished, we were delighted to receive the news that an article on purpose in biology that we had written for the oldest biological society in the world, the Linnean Society, had been accepted for publication. We cannot do better than to finish this book just as we finished that article:

It will require creative ingenuity to shift the culture of biology away from the misunderstandings of the twentieth century. If we date the dogmatic hardening of the Modern Synthesis as 1970, then that misleading culture has embedded itself for half a century. We cannot suddenly recreate the pre-1970s culture when integrative functional biology experienced many golden periods of discovery. It will be for a new generation to discover and create their own culture fit for the challenges of the twenty-first century.

They will have plenty of looming signposts to warn them what went wrong. Theirs will be a generation that must take responsibility for the way in which the earth's ecosystems need rescuing, even for our own species to survive. Theirs will be the generation that faces the challenge of aging societies, requiring medical science to find solutions to diseases of old age that do not readily yield to reductionist gene-centric solutions since those diseases are multifactorial. Only an integrative approach that understands those multifactorial interactions can possibly hope to address those diseases.

Theirs will be a generation that can try to recover from the damage to society that results from reductionist models of physiology and evolution that have metaphorically shaped ideas and models in fields as diverse as economics, sociology, philosophy, ethics, politics ... the list goes on because no aspect of today's society can have escaped dogmas like 'we are born selfish', 'they [genes] created us body and mind', 'it's in their DNA', and the myriad of other tropes of related types that we now use almost without thinking.

Those future generations will also need to rewrite the textbooks, not only because they see the virtue of 'let us therefore teach our children', but also because their politicians, economists, sociologists and philosophers will also need to find new strategies, in collaboration with biologists who can lead them out of the gene-centric impasse.

It is arguably a challenge the scale of which human society has never faced before. We wish them all well.

Summary of Common Misunderstandings

Changes in the structure and function of organisms in one generation cannot be passed on through the germ line. This dogma was formulated by August Weissman in 1883 and, in the mid-twentieth century, became a fundamental part of a gene-centric dogma. Weismann provided no evidence at the time for his 'barrier' and we now know that the barrier is a permeable boundary through which the organism can influence its germ line.

The organism cannot alter its genes, so causation is held to be a one-way process from gene to organism functionality. On this view, the genome alone is sufficient to characterise a living organism. We show that this cannot be true. Much more than the genome is inherited, and the genome is interpreted by the organism in reaction with its environment, consisting largely of other living organisms.

The organism is a passive vehicle for retaining genes in a 'gene pool' and, most significant, the behaviour and function of organisms is controlled to this end. This gave birth to the selfish-gene concept, popularised by Richard Dawkins in his best-selling book, *The Selfish Gene*. We show that selfish and cooperative behaviours depend on the organism and its social interactions with other living organisms, not on the organism's genes.

Evolution occurs through small random changes in genes (gene mutation) that are passively selected in the process of natural selection. We show that there are many other processes operating in the evolution of living organisms.

Living organisms are not agents, but rather passively experience random change. We show that living organisms are necessarily purposive agents, and that they can influence their own and their species' development and evolution.

References and Further Reading

Chapter 1

For a general introduction to genes, see Kampourakis, K. 2021. *Understanding Genes*. Cambridge: Cambridge University Press.

Mendel himself did not talk about 'genes'. That word was introduced in the early twentieth century. What we mean by 'Mendelian gene' is the inheritable trait that he was studying, such as pea shape or colour. See Hartl, D. L. & Orel, V. 1992. What did Gregor Mendel think he had discovered? *Genetics* 131, 245–253; van Dijk, P. & Ellis, T. H. N. 2016. The full breadth of Mendel's genetics. *Genetics* 204, 1327–1336. https://doi.org/10.1534/genetics .116.196626; Auffray, C. & Noble, D. 2022. Gregor Mendel at the source of genetics and systems biology. *Biological Journal of the Linnean Society* 137, 720–736. https://doi.org/10.1093/biolinnean/blac105.

Dawkins, R. 1976. *The Selfish Gene*. Oxford: Oxford University Press.

On the inadequacy of Dawkins' concept and its slipperiness, see Noble, D. 2008. Genes and causation. *Philosophical Transactions of the Royal Society A*, 366, 3001–3015; Noble, D. 2011. Neo-Darwinism, the Modern Synthesis and selfish genes: are they of use in physiology? *Journal of Physiology* 589, 1007–1015; Noble, D. & Noble, R. 2022. The origins and demise of selfish gene theory. *Theoretical Biology Forum* 115, 29–43. https://doi.org/10.19272/202211402003.

A transcript of the debate between Lynn Margulis, the originator of symbiogenesis in evolution, and Richard Dawkins can be accessed at https://www.denisnoble.com /wp-content/uploads/2019/11/HOMAGE_COMMENTARY_Music-of-Life-1.pdf. The relevant part of the transcript is on page 24.

For an entertaining and informative perspective on the issue of genes as programming life, see Coen, E. 2000. *The Art of Genes: How Organisms Make Themselves*. Oxford: Oxford University Press. 'Organisms are not simply manufactured according to a set of instructions. There is no way to separate instructions from the process of carrying them out, to distinguish plan from execution.'

For how organisms make choices, see Noble, R. & Noble, D. 2018. Harnessing stochasticity: how do organisms make choices? *Chaos* 28, 106309. https://doi .org/10.1063/1.5039668.

For a standard evolutionary biology perspective on the transitions during evolution, see Maynard Smith, J. & Szathmary, E. 1998. *The Major Transitions in Evolution*. Oxford: Oxford University Press. Our extensions of this approach to evolution are described in Noble, R. & Noble, D. 2022. Physiology restores purpose to evolutionary biology. *Biological Journal of the Linnean Society*. https://doi.org /10.1093/biolinnean/blac049.

Early work on reconstructing the cardiac pacemaker can be found in Noble, D. 2006. *The Music of Life*. Oxford: Oxford University Press, Chapter 5, pp. 55–73. For a more detailed account of the research on which the section on the heartbeat is based, see Noble, D. 2021. The surprising heart revisited: an early history of the funny current with modern lessons. *Progress in Biophysics and Molecular Biology* 166, 3–11.

Calculations of what is transmitted to future generations from the complete cell in addition to the genome show that this is vastly greater than that transmitted by the genome alone. See e.g. Noble, D. 2017. Digital and analogue information in organisms. In Walker, S. I., Davies, P. C. W. & Ellis, G. F. R. (eds.), *From Matter to Life: Information and Causality*. Cambridge: Cambridge University Press, pp. 114–129.

In the early days of molecular biology, it was thought that each molecular gene was responsible for just one protein in a biochemical pathway (Beadle, G. W. & Tatum, E. L. 1941. Genetic control of biochemical reactions in Neurospora, *PNAS* 27, 499–506). Later it became evident that the relation between a gene, meaning a specific DNA sequence, and biological function is much more complex. Many biologists now favour the omnigenic hypothesis, which proposes that all genes are involved in all biological functions (Boyle, E. A., Yang, I. L. & Pritchard, K. P. 2017. An expanded view of complex traits: from polygenic to omnigenic. *Cell* 169, 1177–1186). The view of the authors is that there are no 'genes for' this and that. See Noble, D. 2016. *Dance to the Tune of Life: Biological Relativity*. Cambridge: Cambridge University Press, pp. 138–152 on 'The language of neo-Darwinism'.

Chapter 2

Schrödinger, E. 1944. *What is Life? The Physical Aspect of the Living Cell*. Cambridge: Cambridge University Press.

The first person to notice the mistake in Schrödinger's work was Jean-Jacques Kupiec in his 2009 book *The Origin of Individuals* (World Scientific Publishing), based on work originally published in French in 1981: Theorie probabilistic de la differentiation cellulaire. *XIIème Rencontres de Méribel*, 161–163. The details of the argument against Schrödinger can be found in Noble, D. 2016. *Dance to the Tune of Life: Biological Relativity*. Cambridge: Cambridge University Press, pp. 134–136, and in more detail in Noble, D. 2021. The illusions of the Modern Synthesis. *Biosemiotics* 14, 5–24. https://doi.org/10.1007/s12304-021-09405-3 (see section entitled 'Illusion 4. The illusion of the Central Dogma').

The Francis Crick quotations are from Crick, F. 1970. Central dogma of molecular biology. *Nature* 227, 561–563. https://doi.org/10.1038/227561a0; and Crick, F. 1988. *What Mad Pursuit: A Personal View of Scientific Discovery*. New York: Basic Books.

Watson, J. D. & Crick, F. H. C. 1953. Molecular structure of nucleic acids: a structure for deoxyribose nucleic acid. *Nature* 171, 737–738.

Sydney Brenner contributed some of the important steps in understanding the DNA triplets and how these are used in specifying the amino acid sequence of a protein. He wrote *A Life in Science*, published by Biomed Central in 2001.

Noble, D. 2006. *The Music of Life. Biology beyond Genes*. Oxford: OUP. Especially pp. 42–54 on why the genome is not a program.

The evidence for the failure of standard texts of evolutionary biology to include many major discoveries has been extensively documented by Shapiro, J. A. & Noble, D. 2021. What prevents mainstream evolutionists teaching the whole truth about how genomes evolve? *Progress in Biophysics and Molecular Biology* 165, 140–152. This article documents 40 major discoveries in evolutionary biology during the twentieth century, only three of which are to be found in standard textbooks and popularisations of evolutionary biology.

Barbara McClintock's Nobel Prize lecture was published in *Science*: McClintock, B. 1984. The significance of responses of the genome to challenge. *Science* 226, 792–801.

A second, and much expanded, edition of Shapiro's book has now appeared: Shapiro, J. A. 2022. *Evolution: A View From the 21st Century. Why Evolution Works as Well as it Does*. Chicago: Cognition Press. With over 600 pages, this book has become a massive literature source for all the forms of editing of genomes by organisms. Shapiro coined the phrase *read–write genome* for this property, as opposed to the *read-only* view of neo-Darwinism and the Central Dogma.

Evidence that massive reorganisation of genomes has occurred during the evolution of different species comes from the *Nature* paper in 2001 announcing the full sequencing of the human genome: International Human Genome Sequencing Consortium (Lander, E. S. & many others). 2001. Initial sequencing and analysis of the human genome. *Nature* 409, 860–921. https://doi.org/10.1038/35057062. Figure 42 of this paper shows the evidence that whole domains of chromatin and transcription-factor proteins have been recombined as various species from yeast to human have evolved. A description in more accessible language can be found in Noble, D. 2016. *Dance to the Tune of Life: Biological Relativity*. Cambridge: Cambridge University Press, pp. 201–204 and Figure 7.4.

Anthwal, N., Joshi, L. & Tucker, A. S. 2013. Evolution of the mammalian middle ear and jaw: adaptations and novel structures. *Journal of Anatomy* 222, 147–160. https://doi.org/10.1111/j.1469-7580.2012.01526.x: 'Having three ossicles in the middle ear is one of the defining features of mammals. All reptiles and birds have only one middle ear ossicle, the stapes or columella. How these two additional ossicles came to reside and function in the middle ear of mammals has been studied for the last 200 years and represents one of the classic examples of how structures can change during evolution to function in new and novel ways.'

The incorrect idea that the Weismann barrier idea is supported by the Central Dogma of molecular biology is widespread on the internet, for example on Wikipedia. See Noble, D. 2018. Central Dogma or Central Debate? *Physiology* 33, 246–249.

Physiological evidence for epigenetic inheritance is rapidly growing. The review from which the quotation comes is Sales, V. M., Ferguson-Smith, A. C. & Patti, M.-E. 2017. Epigenetic mechanisms of transmission of metabolic disease across generations. *Cell Metabolism* 25, 559–571. https://doi.org/10.1016/j .cmet.2017.02.016; see also Danchin, E., Pocheville, A., Rey, O., Pujol, B. & Blanchet, S. 2019. Epigenetically facilitated mutational assimilation: epigenetics as a hub within the inclusive evolutionary synthesis. *Biological Reviews* 94, 259–282. https://doi.org/10.1111/brv.12453.

The story of high- and low-altitude birds and how they have adapted haemoglobins through disparate genome changes can be found in Natarajan, C., Hoffmann, F. G., Weber, R. E. *et al.* 2016. Predictable convergence in hemoglobin function has unpredictable molecular underpinnings. *Science* 354, 336–339. For humans, see Bigham, A. W. & Lee, F. S. 2014. Human high-altitude adaptation: forward genetics meets the HIF pathway. *Genes and Development* 28, 2189–2204.

Chapter 3

The full title of Darwin's first book is *On the Origin of Species by Means of Natural Selection, or the Preservation of Favoured Races in the Struggle for Life*. The quotation is from page 5.

There are many misunderstandings concerning Charles Darwin and his iconic book *The Origin of Species*. From the very beginning, the first edition in 1859, Darwin covered much more than natural selection, since even that first edition included his Lamarckian ideas on the inheritance of use and disuse. Modern neo-Darwinism is not truly Darwinian, therefore, since it excludes precisely these Lamarckian ideas which, in later years, became stronger in Darwin's thought rather than weaker. Neo-Darwinism owes its origins to two nineteenth-century scientists: Alfred Russel Wallace, who opposed Darwin on the idea that organisms could exercise voluntary choice, and August Weismann, who invented the idea of the Weismann barrier to exclude the inheritance of acquired characteristics.

For details of the giant panda story, see Guang, X., Lan, T., Wan, Q. H. *et al.* 2021. Chromosome-scale genomes provide new insights into subspecies divergence and evolutionary characteristics of the giant panda. *Science Bulletin (Beijing)* 66, 2002–2013. https://doi.org/10.1016/j.scib.2021.02.002.

We owe our current knowledge of symbiogenesis to the work of Lynn Margulis: Margulis, L. 1970. *Origin of Eukaryotic Cells*. New Haven, CT: Yale University Press. She took part in a seminal debate with Richard Dawkins in 2009 (https://player.vimeo.com/video/52530866). Lynn Margulis died in 2011. She had succeeded in showing convincing evidence that mitochondria in animal cells had originated from fusion between archaebacterial cells and the ancestors of eukaryotes, leading to much greater generation of ATP from metabolic processes. Her work is celebrated in a documentary film, *Symbiotic Earth*, produced by Hummingbird films: https://hummingbirdfilms .com/symbioticearth.

The first book referring to evolutionary transitions in its title is Maynard Smith, J. & Szathmary, E. 1995 (1998 paperback). *The Major Transitions in Evolution*. Oxford: Oxford University Press. They identify five major transitions: (1) the origin of chromosomes, (2) the origin of eukaryotes, (3) the origin of sex, (4) the origin of multicellular organisms, (5) the origin of social groups. Their list significantly avoids addressing the origins of nervous systems, of associative learning, and of active agency. There must also have been major transitions before the origin of chromosomes.

For the Galapagos finch story, see Lamichhaney, S., Han, F., Webster, M. T., Andersson, L., Grant, B. R. & Grant, P. R. 2018. Rapid hybrid speciation in Darwin's finches. *Science* 359, 224–228. doi: 10.1126/science.aao4593.

The idea that organisms can use the harnessing of stochasticity to feel their evolutionary way forward is in Noble, R. & Noble, D. 2017. Was the watchmaker blind? Or was she one-eyed? *Biology* 6, 47.

The nature of cell membranes rarely features in texts and popularisations of general biology, including evolution. This could be one reason why the inheritance of the membranous structure of cells is downplayed in modern evolutionary biology. Yet there are no genes 'for' phospholipids. Cell membranes are also better candidates for self-replication than DNA since they really do grow like crystals, i.e. by natural accretion of new molecules inserting themselves into the existing structure. The best introduction to cell membranes and their functions is to be found in physiological textbooks. We recommend the sections on 'Structure of biological membranes', 'Function of membrane proteins', and 'Cellular organelles and the cytoskeleton' in Boron, W. & Boulpaep, E. 2016. *Medical Physiology*, 3rd edition. Philadelphia, PA: Elsevier.

The formation of lipid vesicles, called liposomes, was first shown by Alec Bangham in 1961 while working at the Babraham Research Institute near Cambridge. His images using an electron microscope were the first direct evidence for lipid bilayers forming cell membranes.

The technical term for self-maintaining systems is autopoiesis, from the Greek for 'self-creating'. The idea was developed by Humberto Maturana and Francisco Varela in their 1972 book *Autopoiesis and Cognition: The Realization of the Living*. The third edition in English is published by Springer (1991). A good modern textbook on the systems approach to life is Capra, F. & Luisi, P. L. 2014. *The Systems View of Life: A Unifying Vision*. Cambridge: Cambridge University Press.

A good account of the origins of cells and their structures, including the nucleus, is in Harold, F. M. 2014. *In Search of Cell History*. Chicago, IL: University of Chicago Press.

A review of the process called quorum sensing is in Miller, M. B. & Bassler, B. L. 2001. Quorum sensing in bacteria. *Annual Reviews of Microbiology* 55, 165–199.

The constraints imposed on lower levels of biological organisation by higher levels of biological organisation are a central feature of the principle of biological relativity, first published in book form by Noble, D. 2016. *Dance to the Tune of Life: Biological Relativity*. Cambridge: Cambridge University Press.

On the evolution of multicellularity see Miller, S. M. 2010. Volvox, Chlamydomonas, and the evolution of multicellularity. *Nature Education* 3(9), 65.

The quotation from Anton van Leeuwenhoek is from a letter to the Royal Society, 1674. For a detailed account of his discoveries, see Krutch, J. W. 2005. *The Great Chain of Life*. Iowa City, IA: University of Iowa Press, Chapter 2, Machinery for evolution. The book was first published in 1957.

On the remarkable behaviour of cephalopods (octopus and squid, and the group strategies of many other species), see Packard, A. & Dellafield-Butt, J. 2014. Feelings as agents of selection: putting Charles Darwin back into (extended neo-) Darwinism. *Biological Journal of the Linnean Society* 112, 332–353.

Chapter 4

For some of the common misunderstandings on genes and development, see Minelli, A. 2021. *Understanding Development*. Cambridge: Cambridge University Press. Minelli writes (p. 154): 'in the genome there are not specific genes for specific phenotypic traits or genes responsible for the production of a major organ such as heart or eye, and there would be no gene expression were it not for the whole machinery of the cell. Furthermore, all developmental processes take place in a specific environmental context that influences the expression of genes.'

For a more technical treatment of the restraints and other interactions between different levels of organisation see Noble, R. Tasaki, K., Noble, P. J. & Noble, D. 2019. Biological relativity requires circular causality but not symmetry of causation. So, where, what and when are the boundaries? *Frontiers in Physiology* 10, 827. https://doi.org/10.3389/fphys.2019.00827.

The image of chimpanzees using stone hammers for cracking nuts was published in *The Times* (London) in 2020 under the article title 'Chimp culture to be protected by UN'. The reason given is that this culture is not universal to chimpanzee groups in Africa. It is a learnt culture: www.thetimes.co.uk/article/chimp-culture-to-be-protected-by-un-v9lftcnkl.

Heyes, C. 2018. *Cognitive Gadgets: The Cultural Evolution of Thinking*. Cambridge, MA: Harvard University Press.

For the work on the evolution of consciousness see the ground-breaking book, Ginsburg, S. & Jablonka, E. 2019. *The Evolution of the Sensitive Soul: Learning and the Origins of Consciousness*. Cambridge, MA: MIT Press. Their main conclusion is that the origin of consciousness may be as far back as the Cambrian explosion, *c*. 500 million years ago. Consciousness may be much more widespread than commonly assumed.

On the role of purpose in organisms see Noble, R. & Noble, D. 2022. Physiology restores purpose to evolutionary biology. *Biological Journal of the Linnean Society*. https://doi.org/10.1093/biolinnean/blac049.

The Modern Synthesis was originally defined in Julian Huxley's 1942 book, *Evolution: The Modern Synthesis*. Whilst adhering to the neo-Darwinist line of excluding agency and the inheritance of acquired characteristics, Huxley included a much wider range of processes than adherents of the Modern

Synthesis accept today. The hardened, narrowed-down version developed in the 1963 2nd edition, following Crick's formulation of the Central Dogma of molecular biology. The history of this unfortunate development is described in Noble, D. & Noble, R. 2023. How purposive agency became banned from evolutionary biology. In: Corning PA, Kauffman SA, Noble D, Shapiro JA, Vane-Wright RI, Pross A, eds. Evolution "on purpose". Cambridge, MA: MIT Press, 221–235.

For work on Lamarckian forms of evolution, see Gissis, S. B. & Jablonka, E. (eds.) 2011. *Transformations of Lamarckism: From Subtle Fluids to Molecular Biology*. Cambridge, MA: MIT Press; Jablonka, E. & Lamb, M. J. 2005. *Evolution in Four Dimensions: Genetic, Epigenetic, Behavioral, and Symbolic Variation in the History of Life*. Cambridge, MA: MIT Press; Tollefsbol, T. O. (ed.) 2019. *Transgenerational Epigenetics: Evidence and Debate*, 2nd edition. London: Academic Press; Allis, C. D., Caparros, M.-L., Jenuwein, T., Reinberg, D. & Lachner, M. (eds.) 2015. *Epigenetics*, 2nd edition. Cold Spring Harbor, NY: Cold Spring Harbor Laboratory Press.

For work on the evolution of eyes, see Williams, D. L. 2016. Light and the evolution of vision. *Eye* 30, 173–178. https://doi.org/10.1038/eye.2015.220.

On the gut, see Enders, G. 2017. *Gut: The Inside Story of Our Body's Most Underrated Organ*. London: Scribe.

For the history of selfish gene theory, see Noble, D. & Noble, R. 2022. Origins and demise of selfish gene theory. *Theoretical Biology Forum*. 115, 29–43. https://doi.org/10.19272/202211402003.

On tree migration see St George, Z. 2020. *The Journeys of Trees: A Story about Forests, People, and the Future*. New York: W. W. Norton.

On plant behaviour see Trewavas, A. 2015. *Plant Behaviour and Intelligence*. Oxford: Oxford University Press.

Chapter 5

On the role of boundaries in physiology, see Noble, R. Tasaki, K., Noble, P. J. & Noble, D. 2019. Biological relativity requires circular causality but not symmetry of causation. So, where, what and when are the boundaries? *Frontiers in Physiology* 10, 827. https://doi.org/10.3389/fphys.2019.00827.

Burgess, M. 1990. *The Cry of the Wolf*. London: Andersen Press.

On the behaviour of wolves, see Bekoff, M. & Pierce, J. 2009. *Wild Justice: The Moral Lives of Animals*. Chicago, IL: University of Chicago Press.

On hearing in barn owls, see Krings, M., Rosskamp, L. & Wagner, H. 2018. Development of ear asymmetry in the American barn owl (*Tyto furcata pratincola*). *Zoology (Jena)* 126, 82–88. https://doi.org/10.1016/j .zool.2017.11.010.

On the dodder plant, see Shen, G., Liu, N., Zhang, J., Xu, Y., Baldwin, I. T. & Wu, J. 2020. *Cuscuta australis* (dodder) parasite eavesdrops on the host plants' FT signals to flower. *PNAS* 117, 23125–23130. https://doi.org/10.1073/pnas .2009445117.

On the inheritance of stroking behaviour, see Weaver, I. C. G. 2009. Life at the interface between a dynamic environment and a fixed genome. In Janigro, D. (ed.), *Mammalian Brian Development*. New York: Humana Press, pp. 17–40; Champagne, F. A., Weaver, I. C. G., Diorio, J., Dymov, S., Szyf, M. & Meaney, M. J. 2006. Maternal care associated with methylation of the estrogen receptor-alpha1b promoter and estrogen receptor-alpha expression in the medial preoptic area of female offspring. *Endocrinology* 147, 2909–2915.

On behavioural choice in monkeys, see Brosnan S. F. & de Waal, F. B. M. 2003. Monkeys reject unequal pay. *Nature* 425, 297–299.

On behavioural choice in dogs, see Essler, J. L., Marshall-Pescini, S. & Range, F. 2017. Domestication does not explain the presence of inequity aversion in dogs. *Current Biology* 27, 1861–1865.

On the physiology of intentional choice, see Noble, R. & Noble, D. 2021. Can reasons and values influence action: how might intentional agency work physiologically? *Journal of the General Philosophy of Science* 52, 277–295. https://doi.org/10.1007/s10838-020-09525-3.

Chapter 6

Watson, J. B. 1913. Psychology as the behaviorist views it. *Psychological Review* 20, 158–177.

For information on the Skinner box, also called the operant conditioning box: McLeod, S. 2018. What is operant conditioning and how does it work? How reinforcement and punishment modify behavior. *Simply Psychology*. www .simplypsychology.org/operant-conditioning.html.

Understanding the forest: on tree migration, see St George, Z. 2020. *The Journeys of Trees: A Story about Forests, People, and the Future*. New York: W. W. Norton.

Charles Darwin clearly understood choice in animals. In his 1871 book *The Descent of Man and Selection in Relation to Sex* he wrote 'When we behold two males fighting for the possession of the female, or several male birds displaying their gorgeous plumage, and performing strange antics before an assembled body of females, we cannot doubt that, though led by instinct, they know what they are about, and consciously exert their mental and bodily powers' (p. 245). This paragraph is significant because he clearly acknowledges the role of inherited tendencies (instincts) but sees them as overlain by conscious intent.

The 2019 book *The Evolution of the Sensitive Soul*, by Simona Ginsburg and Eva Jablonka, is based on first developing a criterion for conscious agency, which is unlimited associative learning, and then exhaustively showing which organisms display this ability *as a matter of experimental observation*. The conclusion of this scientific empirical study is that the origins of this ability must go back at least to the time of the Cambrian explosion.

For experimental evidence that animals can make moral choices, see Brosnan S. F. & de Waal, F. B. M. 2003. Monkeys reject unequal pay. *Nature* 425, 297–299; Essler, J. L., Marshall-Pescini, S. & Range, F. 2017. Domestication does not explain the presence of inequity aversion in dogs. *Current Biology* 27, 1861–1865.

On foxes: McDonald, D. 1987. *Running with the Fox*. London: Unwin Hyman.

Chapter 7

The special issue of the *RUSI Journal* is volume 164, issue 5–6. The issue ends with a forum: De Angelis, E., Hossaini, A., Noble, D. *et al.* 2019. Artificial intelligence, artificial agency and artificial life. *Journal of the Royal United Services Institute* 164, 120–144.

Stochastic movement of particles in water was first observed by studying the movements of ground pollen particles using a microscope by Robert Brown in 1827. The phenomenon was explained in 1905 by Albert Einstein, who showed that it was due to the random bombardment of suspended grains by water molecules. One of the major mistakes of twentieth-century biology was to ignore the stochasticity of water-based living systems, and to suppose instead that they arise from a determinate read-out of crystalline genetic material. Contrary to this view, we have developed the principle of the harnessing of stochasticity, since all biological systems are water-based and must display this phenomenon.

For a review of how cloning of a sheep, and other animals, was achieved, see Niemann, H., Tian, X. C., King, W. A. & Lee, R. S. F. 2008. Epigenetic reprogramming in embryonic and foetal development upon somatic cell nuclear transfer cloning. *Reproduction* 135, 151–163. https://doi.org/10.1530/REP-07-0397.

On the links between poverty, intrauterine development and disease, see Poston, L., Godfrey, K. M., Gluckman, P. D. & Hanson, M. A. 2023. *Developmental Origins of Health and Disease*, 2nd edition, Cambridge: Cambridge University Press.

Chapter 8

Boysen Jensen, P. 1910. Über die Leitung des phototropischen Reizes in Avenakeimpflanzen. [On the conduction of the phototropic stimulus in Avena plants.] *Berichte der Deutschen Botanischen Gesellschaft* 28, 118–120.

Plants as timekeepers: Sanchez, S. E. & Kay, S. A. 2016. The plant circadian clock: from a simple timekeeper to a complex developmental manager. *Cold Spring Harbor Perspectives in Biology* 8(12), a027748. https://doi.org/10.1101/cshperspect.a027748.

The principle of biological relativity was first formulated in Noble, D. 2012. A theory of biological relativity: no privileged level of causation. *Interface Focus* 2, 55–64. The physics and biology backgrounds to the principle are analysed in Noble, D. 2016. *Dance to the Tune of Life: Biological Relativity*. Cambridge: Cambridge University Press. Chapter 1 gives the physics background, and Chapters 2–4 the biology background. Chapter 6 describes the principle. Chapter 7–9 give the consequences for evolutionary biology and epistemology. The principle is further elaborated in Noble, R. Tasaki, K., Noble, P. J. & Noble, D. 2019. Biological relativity requires circular causality but not symmetry of causation. So, where, what and when are the boundaries? *Frontiers in Physiology* 10, 827. https://doi.org/10.3389/fphys.2019.00827.

Heart-rate variability was known to be a sign of good health even 2,000 years ago when the Chinese *Pulse Classic* was written. For a modern treatment see Singh, N., Moneghetti, K. J., Christle, J. W. *et al.* 2018. Heart rate variability: an old metric with new meaning in the era of using health technologies for health and exercise training guidance. Part One: physiology and methods. *Arrhythmia and Electrophysiology Review* 193–198. https://doi.org/10.15420/aer.2018.27.2.

On lions and wildebeest: Fryxell, J., Mosser, A., Sinclair, A. & Packer, C. 2007. Group formation stabilizes predator–prey dynamics. *Nature* 449, 1041–1043. https://doi.org/10.1038/nature06177.

On hermit crabs: www.discovermagazine.com/planet-earth/hermit-crabs-line-up-by-size-to-exchange-shells.

On yellow-necked mice: www.woodlandtrust.org.uk/trees-woods-and-wildlife/animals/mammals/yellow-necked-mouse.

On sociable weavers: www.audubon.org/magazine/march-april-2014/africas-social-weaverbirds-take-communal.

On palolo worms: Burrows, W. 1945. Periodic spawning of 'palolo' worms in Pacific waters. *Nature* 155, 47–48. https://doi.org/10.1038/155047a0.

On whales talking: www.nationalgeographic.com/animals/article/scientists-plan-to-use-ai-to-try-to-decode-the-language-of-whales.

Chapter 9

For the physiological and philosophical background of this chapter, on the physiology of intentional choice, see Noble, R. & Noble, D. 2021. Can reasons and values influence action: how might intentional agency work physiologically? *Journal of the General Philosophy of Science* 52, 277–295. https://doi.org/10.1007/s10838-020-09525-3.

On Einstein's brain: Balter, M. 2009. Closer look at Einstein's brain: a new anatomical study reveals unusual features in the physicist's parietal lobes. www .science.org/content/article/closer-look-einsteins-brain. The original article in *The Lancet* is Witelson, S. F., Kigar, D. L. & Harvey, T. 1999. The exceptional brain of Albert Einstein. *Lancet* 353, 2149–2153. https://doi.org/10.1016/S014 0-6736(98)10327-6.

The quotation from Einstein is commonly attributed to him, and is thought to have originally been made in 1945.

On orangutans: Orangutans. *National Geographic*. www.nationalgeographic.com /animals/mammals/facts/orangutans.

Paton Walsh, J. 1994. *Knowledge of Angels*. Cambridge: Green Bay.

For the real-life cases of feral children, see Macdonald, F. 2015. Feral: the children raised by wolves. *BBC Culture*. www.bbc.com/culture/article/20151012-feral-the-children-raised-by-wolves. The cases described are based on real examples of feral children, but the images of them as feral, living amongst animals, have been reconstructed from their stories and from the information on how they were discovered and what kind of animals they were living with.

For the physiological studies of twins, see Keul, J., Dickhuth, H. H., Simon, G. & Lehmann, M. 1981. Effect of static and dynamic exercise on heart volume, contractility, and left ventricular dimensions. *Circulation Research* 48, 1162–1170; Bathgate, K. E., Bagley, J. R., Jo, E. *et al.* 2018. Muscle health and performance in monozygotic twins with 30 years of discordant exercise habits. *European Journal of Applied Physiology* 118, 2097–2110. https://doi.org/10 .1007/s00421-018-3943-7.

For the article from which the ending paragraphs of the book were taken, on the role of purpose in organisms, see Noble, R. & Noble, D. 2022. Physiology restores purpose to evolutionary biology. *Biological Journal of the Linnean Society*. https://doi.org/10.1093/biolinnean/blac049.

Figure and Quotation Credits

Figures

Figure 2.1   Reproduced from https://www.genome.gov/sites/default/files/tg/en/illustration/double_helix.jpg. Image in public domain.

Figure 3.1   Reproduced from Guang, X., Lan, T., Wan, Q. H. *et al.* 2021. Chromosome-scale genomes provide new insights into subspecies divergence and evolutionary characteristics of the giant panda. *Science Bulletin (Beijing)* 66, 2002–2013. Reproduced with permission from Elsevier. © 2021 Science China Press. Published by Elsevier B.V. and Science China Press.

Figure 3.2   Reproduced from Noble, R. & Noble, D. 2022. Physiology restores purpose to evolutionary biology. *Biological Journal of the Linnean Society*. https://doi.org/10.1093/biolinnean/blac049. Made available under a Creative Commons CC BY 4.0 licence (https://creativecommons.org/licenses/by/4.0). © 2022 The Linnean Society of London.

Figure 4.2   Clive Bromhall / Photodisc / Getty Images.

Figure 6.1   Reproduced from https://commons.wikimedia.org/wiki/File:Skinner_box_scheme_01.png. Made available under a Creative Commons CC BY-SA 3.0 licence (https://creativecommons.org/licenses/by-sa/3.0).

Figure 8.1   Reproduced from Foster, R. G. & Kreitzman, L. 2014. The rhythms of life: what your body clock means to you! *Experimental Physiology* 99, 599–606. https://doi.org/10.1113/expphysiol.2012.071118. Reproduced with permission from John Wiley and Sons. © 2013 The Authors. *Experimental Physiology* © 2013 The Physiological Society.

Figure 9.1   Elmar Weiss / 500px / Getty Images.

Figure 9.2   Reproduced from Keul, J., Dickhuth, H. H., Simon, G. & Lehmann, M. 1981. Effect of static and dynamic exercise on heart volume, contractility, and left ventricular dimensions. *Circulation Research* 48, 1162–1170. Reproduced with permission. © Wolters Kluwer Health 1981.

Quotations

Chapter 9   Reproduced from Noble, R. & Noble, D. 2022. Physiology restores purpose to evolutionary biology. *Biological Journal of the Linnean Society*. https://doi.org/10.1093/biolinnean/blac049. Made available under a Creative Commons CC BY 4.0 licence   (https://creativecommons.org/licenses/by/4.0).   © 2022 The Linnean Society of London.

Index